UNEVEN DEVELOPMENT AND REGIONALISM: State, Territory and Class in Southern Europe

CROOM HELM SERIES IN GEOGRAPHY AND ENVIRONMENT
Edited by Alan Wilson, Nigel Thrift, Michael Bradford and Edward W. Soja

CONCEPTS AND TECHNIQUES IN URBAN ANALYSIS
Bola Ayeni

URBAN HOUSING PROVISION AND THE DEVELOPMENT PROCESS
David Drakakis-Smith

URBAN TRANSPORT PLANNING
John Black

DAVID HARVEY'S GEOGRAPHY
John L. Paterson

PLANNING IN THE SOVIET UNION
Judith Pallot and Denis J.B. Shaw

CATASTROPHE THEORY AND BIFURCATION
A.G. Wilson

THEMES IN GEOGRAPHIC THOUGHT
Milton E. Harvey and Brian P. Holly

REGIONAL LANDSCAPES AND HUMANISTIC GEOGRAPHY
Edward Relph

CRIME AND ENVIRONMENT
R.N. Davidson

HUMAN MIGRATION
G.J. Lewis

THE GEOGRAPHY OF MULTINATIONALS
Edited by Michael Taylor and Nigel Thrift

URBANISATION AND PLANNING IN THE THIRD WORLD SPATIAL PERCEPTIONS AND PUBLIC PARTICIPATION
Robert B. Potter

OFFICE DEVELOPMENT: A GEOGRAPHICAL ANALYSIS
Michael Bateman

URBAN GEOGRAPHY
David Clark

RETAIL AND COMMERCIAL PLANNING
R.L. Davies

INSTITUTIONS AND GEOGRAPHICAL PATTERNS
Edited by Robin Flowerdew

MATHEMATICAL PROGRAMMING METHODS FOR GEOGRAPHERS AND PLANNERS
James Killen

THE LAND PROBLEM IN THE DEVELOPED ECONOMY
Andrew H. Dawson

GEOGRAPHY SINCE THE SECOND WORLD WAR
Edited by R.J. Johnston and P. Claval

THE GEOGRAPHY OF WESTERN EUROPE
Paul L. Knox

THE GEOGRAPHY OF UNDERDEVELOPMENT
Dean Forbes

REGIONAL RESTRUCTURING UNDER ADVANCED CAPITALISM
Edited by Phil O'Keefe

MULTINATIONALS AND THE RESTRUCTURING OF THE WORLD ECONOMY
Michael Taylor and Nigel Thrift

THE SPATIAL ORGANISATION OF CORPORATIONS
Ian M. Clarke

THE GEOGRAPHY OF ENGLISH POLITICS
R.J. Johnston

WOMEN ATTACHED: THE DAILY LIVES OF WOMEN WITH YOUNG CHILDREN
Jacqueline Tivers

THE GEOGRAPHY OF HEALTH SERVICES IN BRITAIN
Robin Haynes

POLITICS, GEOGRAPHY AND SOCIAL STRATIFICATION
Edited by Keith Hoggart and Eleonore Kofman

PLANNING IN EASTERN EUROPE
Andrew H. Dawson

PLANNING CONTROL: PHILOSOPHIES, PROSPECTS AND PRACTICE
Edited by M.L.Harrison and R. Murdey

UNEVEN DEVELOPMENT AND REGIONALISM

State, Territory and Class in Southern Europe

COSTIS HADJIMICHALIS

CROOM HELM
London • Sydney • Wolfeboro, New Hampshire

Croom Helm Ltd, Provident House, Burrell Row,
Beckenham, Kent, BR3 1AT

Croom Helm Australia, 44-50 Waterloo Road,
North Ryde, 2113, New South Wales

British Library Cataloguing in Publication Data

Hadjimichalis, Costis
Uneven development and regionalism: state,
territory and class in southern Europe.
1. Regional planning — Europe, Southern
2. Europe, Southern — Economic conditions
— Regional disparities
I. Title
330.94 HC244.5

ISBN 0-7099-3700-8

Croom Helm, 27 South Main Street,
Wolfeboro, New Hampshire 03894-2069, USA

Library of Congress Cataloging-in-Publication Data

Hadjimichalis, Costis.
Uneven development and regionalism.

Bibliography: p.
1. Europe, Southern — Economic conditions — Regional
disparities. I. Title.
HC244.5.H33 1986 330.94'0558 86-19649
ISBN 0-7099-3700-8

Printed and bound in Great Britain by Mackays of Chatham Ltd, Kent

CONTENTS

To think history geographically is not contrary to Marxism. It would, however, be more Marxist to think geography historically.

Pierre Vilar

We no longer share that belief in the millenium founded on a few iron laws concerning the inevitability of a democratic-socialist revolution; nor do we enjoy the support of a fatherland of democratic socialism. But one thing is certain: socialism will be democratic or it will not be at all.

Nicos Poulantzas

PREFACE

This book has been a long time in the making. It began as a comparative research in the mid '70s, but its present form is the outcome of the last three years of hectic writing. Thus, I find it difficult to write today about uneven regional development in southern Europe, when much of what was familiar and up-to-date when I started the research has since disappeared or substantially changed. Nevertheless, my basic aim to link geography to the wider economic, social and political structures of southern Europe still holds. Through this interest I have become increasingly convinced of the current political importance of space, something which many in the Left still do not recognize. The geography of southern European societies makes a difference to the way they develop, and this has profound political implications.

Over the years many friends, colleagues and comrades have been generous in ideas and help of various kinds. I recognize how invidious it is to name but a few among many and I know that not all of those I mention will be pleased with the result. I would like to thank L. Dedousopoulos, M. Papayannakis, G. Garofoli, D.G. Ramon, M. Bebil and J. Torras who read various chapters and made helpful suggestions. Special thanks are due to H. Golemis, L. Tsoulouvis, P. Cooke and R. Hudson who read the whole text and whose comments helped me to improve it. This study owes much of its merit to Ed Soja, who has read it several times and has suffered more as a friend than as an editor of the Croom Helm Geography and Environment Series. Unfortunately, none of these friends can be blamed for the final product.

Finally, as I do not have a "wife who could be thanked for solace, patience and self effacement", I thank Dina Vaiou for her valuable contribution throughout this work, as co-researcher and critic. To her this book is dedicated.

University of Thessaloniki Costis Hadjimichalis

UNEVEN DEVELOPMENT AND REGIONALISM:
State, Territory and Class in Southern Europe

Chapter 1
INTRODUCTION

The socialist political struggles currently taking place in southern Europe are creating profound challenges to traditional Marxist dogma and discourse.

How can one continue to speak, for example, of a dichotomy between base and superstructure, when the reorganization and survival of capitalism at the regional and international scale increasingly depend on forms of political and ideological mediation (as in the educational and regional policies of the State and the Mediterranean policy of the EEC) which directly affect the supposed 'laws of motion' that are traditionally considered to determine superstructural events and patterns? And, to come closer to our subject, how can traditional Marxist discourse contend with old and new forms of <u>social movements</u> focusing on urban and regional issues of a clearly anti-capitalist and anti-authoritarian character, but not neatly constructed around specific 'class interests' and purely proletarian initiatives?

The old certainties, the famous 'guarantees of history' and the 'iron laws of motion of the capitalist mode of production' are being strongly questioned and the resultant political uncertainty is accompanied by growing theoretical reappraisal. This is what led Poulantzas, in one of his last interviews before his death, to speak of a 'crisis in traditional Marxism', and Lefebvre to write about 'Marxism exploded'. It should be emphasized, however, that this identity crisis of Marxism is not simply the result of a strengthening of bourgeois dominance. In contrast to the reinvigoration of the Right in other countries, socialist rhetoric is still in good currency in southern Europe. While capitalism itself is undergoing a deep economic crisis, its form of representation and legitimization, as evidenced in recent changes of government in France, Greece and Spain, is being modernized and changed through the growing power of Socialist parties.

How long parliamentary socialism will last and how effective it will be, however, is another question.[1] The rise of socialist political power in southern Europe has not been accompanied by a mass popular movement able to present itself as a hegemonic alternative to the bourgeois establishment. As Poulantzas observed:

> ... it is as though the form of existence of the latter (an alternative socialist movement) had been determined by their function of opposition within a certain capitalist order, and they could not survive the dissolution of that order.

I believe that there exists a series of political, ideological and theoretical obstacles which the socialist forces in southern Europe must overcome if a successful solution is to be found to the present crisis. Furthermore, amidst other important problems, such as the extension of democracy and popular control over politics and the economy, I wish to focus attention on those arising from the existence, significance and consciousness of regional differentiation. In this book, my aim is to contribute to a reinterpretation of uneven regional development in southern Europe through an analysis of the relationship between regionalization (a process designed 'from above' by the state and/or by capital, aiming at restructuring the spatial division of labour according to the changing needs of profitable accumulation) and regionalism (a reaction 'from below' of a specific local social group or a certain social alliance, whose local interests are against such a regionalization).

The relationship between regionalization and regionalism is a trans-historical one, i.e. it is not restricted only to the capitalist mode of production. It takes, however, distinct forms and particular importance in specific capitalist-social formations, when uneven regional development is combined with political and cultural/ideological issues (Hadjimichalis, 1983). Such a case seems to be the current situation in southern Europe, and particularly in Spain, Italy and Greece, which are the concrete frame of reference for this study.[2]

The relationship between regionalization and regionalism and its roots in uneven regional development is viewed as a social confrontation over the spatiality of contemporary capitalism (Soja, 1981). My argument builds upon a conceptualization of spatiality as the material form of the social relations of production, as the concrete historical and territorial framework for accumulation, and hence as the terrain for new forms of class struggle and political mobilization. In this conceptualization, the state, territory and social classes are implanted directly in the analysis of spatiality, not as peripheral categories but as its essential problematic (Tsoukalas, 1981; Cooke, 1983).

Within this general interpretive framework, four more specific objectives can be identified.

First, through an historical and political analysis I will trace the origins and development of regional unevenness in southern Europe, arguing that regional differentiation is the outcome of a spatially differentiated accumulation process and is not due to supposed localized inadequacies of particular regions and people, as it is sometimes claimed. My basic thesis is that uneven regional development is not just the result of the social processes of capitalist development but rather it is also an active moment, in them (to use Harvey's [1982] formulation) helping to shape those processes. Close to this is the need to locate analysis of uneven regional development, not merely within the context and territorial boundaries of a particular national state but rather to make the starting point of such analysis the global character of the world capitalist system.

Secondly, I will try to develop an alternative explanation of uneven regional development by challenging the current understanding of the relationship between the Marxian theory of value, development and space.[3] More specifically, I will try to go beyond immediate appearances of geographical phenomena, by investigating their underlying mechanisms and tendencies. To begin the analysis with such issues as capital mobility, agglomeration in space, spatial divison of labour, development of productive forces, etc., is a necessary, but too elementary an approach. Instead, I will try to use the labour theory of value (LTV) in geographical terms, in a framework dominated until recently by a-spatial categories and processes. In so doing, I propose an alternative framework for possible future research in the field, based on the notion of the geographical transfer of value (GTV). My attention is not restricted to structural forces only (the internal logic of the capitalist mode of production), but equivalent attention is given to the role of social agents in producing and reproducing structures and vice versa. For an illustration of this dual focus, the case of southern European agriculture is analysed in depth.

Thirdly, the development of the previous two points will provide the necessary framework to analyze the relationship between regionalization and regionalism. The particular focus will be on the role of old and new forms of regional social movements, arguing about the political implications of uneven regional development and the possible tactical objectives for planning. In this respect, I will try to avoid the hopeless feeling that nothing can be done, since the iron logic of capitalism destroys any popular initiative. Instead, a more politically optimistic analysis is presented, in which regional planning - together with other aspects of a state intervention - is conceived as an object and arena of struggle.

And fourthly, an attempt will be made to overcome the typical fragmentation and divisive specialization which exists within the relatively small field of regional analysis, in order to reintegrate regional planning with political practice. In doing so, the arguments presented will be critical of both mainstream regional development theories and certain Marxist views and formulations on the 'regional question'.

1.1 REGIONS AND POLITICS IN SOUTHERN EUROPE

Since the middle of the 1960s there has been a growing interest in regional problems throughout Europe, especially in the countries along its southern fringe (Hudson and Lewis, 1984). Mass media and direct state intervention, followed by capital pressures and social reactions from various groups in 'regional peripheries' have introduced regional planning issues alongside other important welfare policies. By the end of the 1970s, regional development became a well established element in corporate, national and supra-national policies (e.g. EEC), while graduate regional studies were booming.

The economic and political circumstances of this period were important in helping to shape the character and content of these new planning initiatives. The 1950s and 1960s were generally marked by high rates of national economic growth, by 'economic miracles' following the long post-war boom. During this period, France, Italy, Spain, Greece and, to a lesser degree, Portugal had the highest annual growth rate in GDP in western Europe, aside from Ireland. In the 1980s, however, such high rates seem to have gone for good. It is not only the international economic crisis, however, which badly affected the economies of southern Europe. More important were the changes in their internal social structure.

The political situation at the beginning of the 1980s was dramatically different from that of the early 1970s. Ten years ago, anti-democratic, largely military, regimes of the reactionary Right, long established (as in Portugal and Spain) or more recent (as in Greece), were in power over much of southern Europe. Except for the reversion to dictatorial military rule in Turkey, this is no longer the case today. Furthermore, PASOK in Greece, PSE in Spain and socialist-led coalition governments in Italy and Portugal soon followed Mitterrand's 1981 election victory in France. For the first time since World War II and the civil wars, 'socialist' governments achieved power throughout southern Europe.

Could we speak then - as Papandreou and Mitterrand declared in 1981 - of a two-tiered Europe, geopolitics conservative in the northern 'core' and socialist in its southern 'periphery'? While this was a popular slogan among these 'socialist' governments, it remained highly problematic, and was not justified in practice. In fact, neither all of the so-called socialist governments, nor their policies were alike. In terms of international relations, there are very diverse attitudes, from Spain and Italy favouring NATO's euro-missile decisions, to France's intervention in Chad, and to Greece's neutralist-tinged anti-Americanism. In terms of domestic policies, with the exception of PSI, all other parties have mobilized sufficient support for electoral victory on the basis of a radical left-looking programme and appealing rhetoric about regional development issues.[4]

But after two or three years in office their programme seems to lack cohesion, and their popular support has started to fluctuate rapidly. Their effort towards modernization and restructuring (especially in the industrial sector) was confronted with popular unrest in spring 1984, when the governments of France and Spain declared thousands of job losses and when the socialist-led Italian government decided to abandon the 'scala mobile' (the automatic adjustment of wages to price indices).[5] The leadership of these neo-socialist governments thus appeared to many to have no real domestic policy other than getting into office (see also, Petras, 1984).

It is perhaps a little unfair to be so critical about each of these parties, especially as the work and attitudes of all of them have collectively helped to develop a new balance of forces in southern Europe. Nevertheless, my criticism is based on the fact that their radical rhetoric about regional development has not yet been translated into an alternative and effective regional practice, especially in the mode of thinking, in institutional reforms and in implementation.[6] What is lacking is a coherent political view of the regional problem. This is not only a failure of socialist parties but also of certain communist parties which, in their avoidance of confrontation, have failed to draw strength from the present crisis of capitalism in southern Europe. In a large part, this failure can be traced back to the impact of state intervention during the last period of 'rightist rule'. Despite their authoritarian character, these earlier interventions changed not only policy instruments and content, but, under the pressure of the masses, they acknowledged regional and local needs in a very direct and concrete way.

At first, this acknowledgment consisted primarily of assistance to 'poor' regions, directly through the improvement of social services and means of collective consumption, as well as indirectly through incentives to capital and labour. But this type of assistance by the end of the 1970s tended to develop - with the help of the rhetoric of the socialist parties - into a much more universal policy statement of what each citizen in each region was entitled to. Thus, the subsequent legitimation of needs has today turned assistance into an obligation, which the state, local authorities and the new 'socialist' leadership have to meet. Virtually all spatial issues have thus become open political questions.

The latter has two effects. Either state obligations generate a popular apathy, with the majority of people 'waiting for solutions' from the state or from short-lived, issue-oriented regional advocacy groups (as in the Greek case). Or, when political conflicts at the local and regional levels do take place, they are substantially convoluted and shunted aside by the inability of the mass-integrative apparatuses (applied as they are in the centralized decisions regarding restructuring) to react to concrete and goal-directed regional interests which are not related to the centralized processes of bargaining and compromise (as in the case of Spain and Italy). Thus, the present 'regional crisis' in southern Europe is not only an acute manifestation of the general pattern of capital restructuring and uneven development in the area but also the manifestation of (a) the concrete antithesis between central state and local community; and (b) of class conflicts which assume a spatial form, when class interests are territorialized.

These observations help to clarify why I am concerned here with 'regions', a term burdened with ambiguity. Regions have become the concrete political arenas in which the effects of both social and spatial marginalization and of state intervention culminate in contemporary southern Europe. Such a view must be differentiated from the 'spatial separatist theme' as has been characterized by Sack (1974) and Gore (1984), that is, the possibility to separate and evaluate the spatial as an independent phenomenon. My extended focus on agriculture, on local social structures and regional social movements will help to clarify that 'regions' do matter in southern Europe. As such, they provide important opportunities for political mobilization and for radical social change.

1.2 THE STRUCTURE OF THE BOOK

The book is structured in two parts. The first is primarily theoretical, although it derives directly from the research presented in the second part. In this respect the theoretical arguments presented here have not been first conceived *in abstracto* and then tested in reality. On the contrary, they are a synthesis of current debates in southern Europe (and elsewhere) and the outcome of a concrete analysis. The challenge, if you like, of this book is to understand the general underlying causes of capitalist unevenness, while at the same time recognizing and appreciating the importance of the specific and the unique.

Thus Chapter 2 criticizes mainstream regional development theories, arguing about their method, their concept of development and their connection with the capitalist state. These theories reproduced themselves in two waves: one carried the pragmatist/conservative view, highlighting free enterprise, free trade and the least possible state intervention; the other carried the liberal/interventionist view, which stressed the necessity of intervention by the welfare state. The chapter ends with a short review of some current regional development approaches, including a critique of neo-liberal theories, promoting 'adjustment' and self-reliance'.

In Chapter 3 an attempt is made to move beyond some theoretical obstacles which cause difficulties in developing a Marxist interpretation of uneven regional development. In this direction, the work of A. Gramsci seems to be particularly helpful, especially his emphasis on civil society, on political and cultural factors and on the state. The core of the chapter includes a critical review of two dominant lines of thought, that of autonomous or semi-autonomous development theories and that of transfer of value development theories. It is argued that both approaches, for different reasons, seem to be inadequate in analysing uneven regional development. A third view is proposed which focuses both upon production and exchange, and upon local social classes and the state. Finally, a short comment on the labour theory of value and the articulation of production and exchange is presented, criticizing both neo-Ricardian views and traditional economistic approaches.

Chapter 4 developing further the third view analyses the two basic theoretical propositions concerning the contradiction of equalization/differentiation and the geographical transfer of value.

In analysing these two propositions, a three level analysis is presented. At the first level, regional analysis is involved with what I propose to call the visible dimensions of uneven development. This level of analysis sometimes confuses the effects with the causes of uneven regional development. At the second level, regional analysis becomes more interested in what I shall call the functional processes that produce existing regional inequalities, emphasizing destruction and disarticulation via mainly 'exogenous' forces. Finally, at the third level, regional analysis is concerned with identifying the underlying tendencies that produce and reproduce uneven regional development, moving beyond the previous two levels but at the same time taking advantage of their findings. A basic thesis of the chapter is that the geographical transfer of value is not an automatic process, an outcome of some internal laws of capital accumulation; it requires the direct and/or indirect intervention of a political agent.

Chapter 5 concludes the theoretical section, providing a synthesis of the previous discussion. The emphasis is shifted to the capitalist state, to the role of local authorities and to the various forms of struggle. A review and critique is presented to instrumentalist, capital-logic, class theories and neo-Gramscian theories of the state. It is shown that, despite their limitations, the last two seem to be the most satisfactory. Attention is also directed towards the role of local authorities, not simply as agents of the central state, but as acquiring a certain specificity of their own, when they are connected with local social interests.

Part II begins with Chapters 6 and 7, where the historical and political context of the problem of uneven regional development in southern Europe is presented. The periodization, with a particular focus on agriculture, identifies four major phases. The first occurs in the 16th and 17th centuries, when Italy and Spain gradually moved from core areas of the world economy to semi-peripheral positions, while Greece remained under Ottoman occupation. The second phase is defined by important internal transformations in social structure (due to industrialization, wars of independence, etc.) and increasing inter-regional and urban-rural differentiation. The third is associated with the development of industrial capitalism, with bourgeois consolidation and popular exclusion, ending with Fascism, civil wars and World War II. Finally, the fourth phase coincides with the post-war era (with important internal subdivisions) when a new international division of labour established itself in southern Europe around the dichotomy and complementarity of 'fordism' and 'peripheral fordism'.

Chapter 8 introduces the problem of agriculture and its connection with uneven regional development. After a critical description of the current situation in the agricultural sector and its deep transformations since the 1950s, the ongoing debate about the appropriate characterization of peasants is presented, together with the political importance of each approach. The chapter ends with a discussion of some theoretical and empirical issues concerning the articulation of peasants with the rest of southern European economy.

Chapters 9 and 10 attempt an empirical analysis, giving an indication of the geographical transfer of value (GTV) in the agricultural sector. The analysis is based on a set of comparative data available for Italy, Spain and Greece over the period 1950 to 1975. The data include gross regional agricultural income and gross regional value added. Evidence about GTV is shown in maps and statistical indices, followed by a first interpretation of existing channels of GTV in agriculture. Finally, the political and ideological mediation of the state and some local institutions is discussed, arguing that the Marxist interpretation of agrarian evolution and GTV cannot be understood in strictly economic terms.

Chapter 11 is concerned with a presentation of certain implications for political practice of the preceding discussion. Two major issues are discussed: authoritarian statism and regional social movements. A categorization of social movements is attempted, based on the distinction between regionalization and regionalism, the existence or not of certain local class alliances and the form and content of each mobilization.

Finally, the concluding Chapter 12 attempts a synthesis, pointing out some of the political implications of the preceding discussion. It also opens some questions for further research concerning new developments in southern Europe.

NOTES

[1] The remarkable losses of socialist votes in 1984 European elections in France, Italy and Spain and, to a lesser degree, in Greece changed the picture of the early 1980s' euphoria and influenced important internal governmental changes in the four countries, see *Le Monde* 30.6.84 and 5.9.84. The most dramatic change was the Rightist victory in the 1986 elections and the growing importance of the Fascist party in France.

[2] The present study is largely based on my research in the late 1970s, when I was writing my PhD. Its present form and content are, however, different from my previous work entitled, *The*

Geographical Transfer of Value: A Comparative Analysis of Regional Development in Southern Europe, PhD., Graduate School of Architecture and Planning, UCLA, 1980. The research was selective in its spatial coverage, for I have concentrated on three countries in southern Europe - Italy, Spain and Greece. This was an outcome of particular knowledge and emotional sympathy, as well as political principle. These three countries (with the addition of Portugal), despite important socio-economic differences, are most closely associated with northern Europe. They experienced Fascism and military dictatorships and have strong leftist movements. My initial aim was to include Portugal in the present study but, due to financial reasons, time and inability to find access to comparative data, this has not been possible.

[3] For specialized references on this subject, see Bauer (1924); Lipietz (1977); Hadjimichalis (1980, 1984); Soja (1981); Harvey (1983): Marelli (1983).

[4] A good example of radical spatial development rhetoric was PASOK's declaration for the 're-organization of Greece's small settlement structure'. The ambitious plan (to be completed in three years!) was founded only on physical planning, while at the same time other ministries and state agencies (as, for example, the Ministry of National Economic Planning, Public Works Administration, the Industrial Bank and others) were promoting different, often contradictory, policies.

[5] See the press of the time, AYGI, 3.4.84 and 5.5.84; Marxism Today, April and May 1984 issues; Le Monde, 10.4.84 and 11.5.84.

[6] In Italy and Spain most institutional reforms concerning regional policies and administration were in fact organized before socialist governments took office. In Greece most declarations for regional reforms remained on paper only, until 1984.

[7] Op. cit (2).

Part I
THEORETICAL FORMULATIONS: THE POLITICAL ECONOMY OF UNEVEN REGIONAL DEVELOPMENT

Historical and political research provides the necessary framework for an understanding of current regional problems. The articulation of different modes of accumulation and labour control in southern Europe from the 16th century to the present corresponded with different social and spatial structures. Since World War II, initially small differences have developed in many European countries into significant regional disparities. And today the regional crisis in southern Europe - as well as in other parts of the globe - attracts attention and generates rhetoric from various sources: from central government bureaucrats and local leaders, from monopoly capital and working class militants, from military and social planners.

It is obvious that differences in political views and perspectives influence scientific regional research. None of these views can be classified as 'ideology free', 'technical', or 'purely scientific' and apolitical. This is because any approach to an aspect of society contains within it, at least implicitly, a view of the whole society. If this view is not static, it contains a view of development and in the last analysis a view of social change. Thus, implicitly or explicitly, different approaches to the regional crisis reflect different philosophies regarding the means, direction and desirability of social change.

In this respect the term 'political economy' is used here to avoid identification with any single academic discipline and to imply an interest in 'the whole' - the central dynamic for social transformation in Marxist thinking. In Part I, I will try to bring together some of the main theoretical and methodological elements required for a Marxist analysis of the political economy of uneven regional development.

The discussion and critique in the following chapters will necessarily be more general than the specific case of southern Europe, because ideas and models of development have evolved and diffused rapidly since World War II and have been applied to various parts of the globe without reservations about their localized applicability. Ultimately, however, my aim is to return to the concrete case of southern Europe in Part II.

The explanation of regionally uneven development in capitalist economies takes many different forms in theory. It follows that action to deal with unevenness in practice, through corresponding regional policies, is dependent on the assumption of particular causes. The following four chapters are primarily concerned with an analysis of these causes and have a dual focus: first, to provide a brief overview and critique of regional development theories, especially those which have been applied in southern Europe; and secondly, to provide a theoretical alternative for a critical understanding of uneven regional development in capitalist societies. The term 'regional' is here, as throughout this book, being used generically for 'spatial', or subnational. For it is clear that the shape of southern Europe's regional map is constantly transformed, so that a pre-defined regionalization would be misleading. 'Regions' and 'localities' are themselves being formed and re-formed, constructed and disintegrated, where the complexity of the geographic mosaic is evident.

Chapter 2

NEW FRONTIERS FOR REGIONAL DEVELOPMENT THEORIES

Until the end of World War II, the term 'development' was used mainly to refer to biological species, real estate and moves in chess. Only thereafter did it become applied to people, countries, sectors and regions. Since then, we have been flooded by development theories whose labels, as I. Illich (1980) observed, are now 'keepsakes for collectors'. We all remember, for example, such catchphrases as import and export substitution, catching-up modernization, dualism, dependency, basic needs, autochthonous industrialization, self-help, etc. Development theorists deposited mountains of reports filled with wishful thinking and simplistic caricatures of opposing approaches. But few could argue today that the results in practice ('out-there') were commensurate with the theorists' efforts and the money spent.

Something which must not be forgotten, however, is that in the immediate post-war period, development experiences were very few. In fact, there were only two, that of Western capitalist economies and the much more recent experience of the Soviet Union. For the majority of societies at that period the range of choice was constrained. The world capitalist system, with its centuries-long established dependencies on the one hand, and the Yalta-shaped East-West Europe on the other, did not allow manoeuvres for Mediterranean countries. Cold war, the Marshall Plan, UNRA and NATO became articulated with local class antagonism to strengthen already dominant groups in capitalist southern Europe. The 'free play of market forces' existed in name only, while almost all development decisions were made on military and political grounds. Development thus became immediately a <u>political question</u>, a tool that 'could stop poverty, hunger, illiteracy and disease that win masses of recruits for socialist revolution' (Marshall Plan, 1947).

When speaking about development theories, therefore, we are in fact speaking primarily about capitalist domination strategies. Capitalist development implies the replacement of general competence and abundant subsistence activites by the specialized production and consumption of commodities. It also implies the gradual monopoly of wage labour over all other work, and generally involves the redefinition of needs in terms of goods and services mass-produced according to expert planning.

These ideas were synthesized and applied in 1951 by an international panel of economic experts, called together by the United Nations. The UN specialists made the now common distinction between developed and underdeveloped countries, regions and sectors and argued that there was only one course open to underdeveloped countries: they should attempt to replicate the recent economic history of the already industrialized capitalist countries. They ought to 'modernize' at all costs.

Regional planning arrived late in this scenario (Friedmann and Weaver, 1979; Weaver, 1982). Although some form of national economic planning has been part and parcel of the capitalist development model from the beginning, the bridge between sectoral planning and spatial planning was lacking. Neoclassical economics did not include an explicit spatial development dimension, while governments at that time did not realize the political cost of uneven regional development. It was not until around 1960 that the regional aspects of economic growth attracted significant attention.[1]

Since then, most European countries have adopted some form of regional planning, and southern Europe was not an exception. In some cases this has merely involved the locational programming of sectoral investments, as in Greece and Portugal. In others, it has entailed the creation of regional development agencies, such as the _Cassa per il Mezzogiorno_ in Italy, charged with the task of accelerating industrial and agricultural development in southern Italy. And in other cases it has meant the adoption of growth pole strategies, such as in Spain and France.

The technical basis for making such decisions, i.e. where resources should be allocated, where urban growth should be encouraged, where particular agricultural production could take place, etc., lies in what may be termed 'regional development theory'. The field is approximately twenty-five years old. Yet, despite its novel appearance and the vast literature and governments' efforts that accompany its evolution, regional development theory is today in crisis. The loss of confidence in the theories of course reflects a wider crisis in the social sciences (Harvey, 1973; Castells, 1983). But this book argues that, within these general trends, the poverty of regional development theory may be traced back to the way in which theorists conceptualize the relationship between 'space', 'development' and 'the state'. To make this argument clear, I will focus upon a set of theories adopted by southern European governments in their efforts towards 'regional integration' of their national territories.

2.1 LABOUR-SURPLUS MODELS AND POLARIZED DEVELOPMENT

Regional development theories appeared in the capitalist tradition and continuously reproduced themselves in two waves. One carried a pragmatist-conservative view, highlighting free enterprise, free trade and the least possible state intervention; the other carried a liberal/ interventionist view, which stressed the necessary intervention by the welfare state, combining, as Jan Tinbergen argued, 'the best elements of capitalism with the best elements of socialism' (Emmerij, 1982, p.12).

The whole history - and I presume the future - of bourgeois regional development theory and practice can be analysed along this antithesis: state versus the market. Although the debate continues today, post-war European regional policy up to the 1980s has been largely a tool of Keynesian state intervention in the process of structural spatial change. As such, it has been a politically responsive, place-oriented welfare policy.

In retrospect it seems that the overwhelming majority of development theories up to the end of the 1960s adopted a variation of what later became known as the labour-surplus model. This model introduced the classical distinction between a modern and a traditional sector. The modern sector (associated with urban-based industrialization, efficient technology and entrepreneurship) was considered to be the locomotive which was to pull the wagons of the traditional sector at a cruising speed. The latter was associated with rural-based, low efficiency agricultural activities, with low 'propensities to save', and thus subordinated to the modern sector. More seriously, the emphasis of policy through allocative planning, whether it was investment or technology or credit or industrialization, was put squarely on the modern, i.e. urban, sector.[2] The main assumption behind it was that the duration of the 'transitional period', during which time the traditional-rural sector 'might' become worse off instead of better, would be acceptable both in political and in human terms. This 'success' model of capitalist development was formulated in depth in 1961 by W. Rostow in The Stages of Economic Growth: A Non-Communist Manifesto. In Rostow's simplified step-model of economic modernization, development was an unidirectional process in which, after a period of transition, a 'take-off' will take place, eventually reaching maturity and sustained prosperity.

This general paradigm of economic development strongly influenced both market-oriented and state interventionist regional development theories. Looking first on the market side, perhaps the most simplistic neoclassical argument is that of scarcity and resource endowment. Thus, backward regions are considered to be in a situation of scarcity, i.e. they lack adequate 'resource endowment'. In contrast, some areas were favoured by climate, minerals, land and water resources, and/or by the ability, intelligence and work ethics of their inhabitants (Perloff and Wingo, 1961). These areas were able to 'take-off' by means of successful exploitation of their resources and mobilization of skills and capital, while others, lacking all or most of the above, remained poor. Richness thus owes a great deal to historical 'accidents' and 'poor luck', as in the case of successful individuals in society,

A related argument has been put forward by marginalist equilibrium theories, assuming free competition, perfect information and perfect mobility of the two 'factors of production', capital and labour. Capitalists maximize profits by locating where marginal costs are lowest, and workers maximize wages by moving to where jobs are to be found. Whatever inequalities exist between and within regional units were considered to be temporary imperfections of the market mechanism. Thus, the whole regional development problem was narrowly defined in theory and related in policy to firm location and labour migration. One aims to arrive at a self-equilibrating situation, where the spatial distribution of income per capita is uniform (Isard, 1960; Hoover, 1967; Alonso, 1967).

A third group of related theories is associated with trade and export-based regional growth (Ohlin, 1933; Borts and Stein, 1962; Williamson, 1965; for the case of Mezzogiorno, see Chenery, 1968). According to these theories, trade and export revenue feeds a process of accumulation and sectoral diversification, leading to a long-run tendency of diminishing regional inequalities. Success of the export base is the determining factor of regional growth, and export-base initiated growth determines employment and income levels in the whole region. The main assumption in this model - which goes back to Ricardo - is that local growth is a function of external demand, so that trade in a market economy proceeds on the basis of comparative advantage and equal exchange.[4]

This group of theories has been - and to some extent still remains - the theoretical cornerstone of regional policies in southern Europe. Under the supervision of European and/or North American advisors, the governments of Portugal, Spain, Italy and Greece have formulated special legislation for resource exploitation, to attract foreign capital and to encourage international migration from backward regions (Nicolinakos, 1973). Resource endowment was the theory used by foreign advisors mainly in Greece and Portugal up to the 1960s. The argument was that both countries were 'poor' in resources, and this could generate regional disequilibria. As B. Ward (1969), a US advisor in Greece, puts it:

> ... In a capital and resources scarce country attempting economic development it is reasonable to expect some concentration in the regional distribution of investment resulting in regional inequalities, because of the concentration of opportunities ... (Ward, 1963, p.57)

The emphasis was put, therefore, primarily on agriculture, which ironically reproduces the same structural inequalities as those it was supposed to overcome. In the case of Greece, this was a blind choice that left underdeveloped its diversified mineral deposits.[5]

Another important policy was to encourage interregional and international migration. In terms of interregional migration, to a considerable extent the effect of such movements has been to convert rural underemployment into urban unemployment, rural into urban poverty (what all this implies in terms of poor housing, living conditions, etc.), although this surplus labour has been important in relation to the growth of industry and services in urban areas (Kalogirou, 1980). Political encouragement of international migration in the 1950s and 1960s coincided with a growing demand for migrant labour in north-western Europe. As with internal migration, there is a considerable selectivity from southern Europe in terms of who moves, where they move to and from, and why they move (Kavouriaris, 1974). This process leads to increasingly underpopulated regions of feminized and aged population, a decline in labour inputs into agriculture and a consequent drop in agricultural output. But this was considered necessary for 'national' economic reasons. As Luis Lefeber, who came to Greece to train government officials in 1966, argues:

> ... Labour in the subsistence regions of Epirus and Thrace has no wages in the competitive sense: its marginal product is zero ... This surplus labour is unproductive and must migrate to urban industrial centers or to commercial agricultural regions or to foreign countries ... (Lefeber, 1966)[6]

In this way, southern European dominant classes succeeded in 'exporting' - at least for a limited period of time - both unemployment and political unrest to western Europe. But short-run positive political effects turned in the 1970s and 1980s into severe development problems. Because, in this way, as Whitman (1970) argues for southern Italy, peripheral regions risk losing their factors of production and at the same time failing to secure second-round exports to more developed regions. In other words, their main exports were more often labour and capital, rather than goods.

Finally, a good deal of controversy arose around the supposed 'structural' and 'locational' disadvantages, endemic in southern Europe. A well known supporter of this thesis was Vera Lutz (1962), who argues that Mezzogiorno suffers from these disadvantages, which render development policies of the type hitherto pursued largely a waste of resources. V. Lutz, following a market-oriented ideology, argues that emigration from the south should be continued to a point at which 'natural' factor proportions will be secured, 'appropriate' to the region's level of development. Apart from a predisposition to a 'natural' development for the south, rather than a 'forced' industrialization, the basis of Lutz' thesis is the claim that the south suffers a locational disadvantage in relation to northern industry. The latter, according to Lutz, is a reflection of higher transport costs and low skills available locally. But, as MIT and SVIMEZ studies have shown, the existence of cheap labour in Mezzogiorno turned their 'disadvantages' to a net labour over transport cost advantage of 3 to 6 per cent. And lack of skills was not sufficient explanation, because companies like IRI and Olivetti have found that it takes only 4 to 6 weeks to train illiterate agricultural labour to perform work necessary for the assembly of automobiles, mechanical office machinery and steel production (Holland, 1971).

During the late 1960s, criticism of these approaches came from various sources, including liberals, monetarists, Marxists and the newly established 'dependency school' (Frank, 1967). Resource endowment, it was argued, can be seen only as an indicator of varying potential for capital accumulation, a potential which is activitated and realized only within particular historical contexts (Amin, 1974: Wallerstein, 1974). Scarcity, therefore, is not something inherent in nature, but its definition is inextricably social and cultural in origin - necessary for the survival of capitalism itself (Harvey, 1974). Similarly, 'structural' and 'locational' disadvantages are not 'natural' phenomena but outcomes of the process of capitalist accumulation, while migration from southern regions was never a 'tradition' but a forced and painful choice. In terms of equilibrium models, the critique depicted their rationalist and reductionist premises, emphasizing the chaotic conception in which space was introduced to an aspatial framework and profit-maximization was restricted by the market size proposed (Massey, 1978b; Holland, 1976; Cooke, 1983). Spatial equilibrium is not an optimum, and it is logically impossible to attain it in a space economy, even when the market mechanism is 'fully lubricated'. As Richardson (1969) has observed, general equilibrium is inconsistent with the implications of the space economy, because it is impossible to have perfect competition when there are transport costs between locations. Finally, a major limitation of export-base models is their total concentration on exports, neglecting both international and national capital flows. Peripheral regions are likely to export multiplier effects and profits along with their exports and fail to allow for the 'spatial inelasticity' in direct capital investment (Lipietz, 1977).

The most celebrated regional policy in southern Europe up to the mid-1970s was the well known growth pole - growth center doctrine. Following the original ideas of Perroux (1955) and Myrdal (1957) about permanent disequilibrium and the necessity of strong state intervention, polarized development models were (and to some extent remain) in good currency around the Mediterranean. The efforts of the _Cassa per il Mezzogiorno_ in Italy, the '_Polos de Desarrollo_' policy in Spain, the French '_métropoles d'équilibre_' and the 'decentralized industrial concentration' of Greece provide examples of these models applied during the 1960s and 1970s. Studies of polarized development produced the still prevailing view that initial stages of capitalist regional development are characterized by increasing inequalities, which decline when more advanced levels are reached. The problem is thus one of time and innovative planning intervention so that the fruits of development will spread out from the modern growth pole down to its traditional periphery.

Although these policies have been in operation for more than two decades, little evidence exists today that their principal goal - to integrate traditional with modern sectors and to diminish regional inequalities - has been met. Industrialization and an increase of exports and per capita income did take place. But the integration was realized primarily between highly capital intensive sectors and in specific locations within the Mediterranean system (and not within southern European regions). This is a common conclusion of both liberal (Lasuen, 1969; Sylos Labini, 1965; Richardson, 1975) and Marxist critics, who emphasized additionally the new International Division of Labour, which shaped the growth poles and maritime complexes in the Mediterranean (Libertini, 1973; Mingione, 1976; Palloix, 1975; Bleitrach and Chenu, 1979; Hadjimichalis, 1976;

Hudson, 1983).

In southern Europe (as well as in other parts of the globe where this strategy has been applied), the modern sector turned into a steel-and-glass enclave with high labour productivity, capital intensive technology and high rates of growth. But it neither acted as locomotive to regional prosperity in social and spatial terms nor created sufficient productive employment. In fact, despite high rates of industrial investment in such regions as Mezzogiorno, southern Spain and northern Greece, out-migration continued until the end of the 1970s. Southern Europe seems to be a good example depicting the confusion around growth poles and its reproduction in planning practice. Regional planners have used the growth pole doctrine to describe a variety of policy interventions which range from port-industrial complexes to small rural villages to serve as service centres. In this sequence, what began as a growing economic unit (in Perrouxian terms), a firm or an industry, has become a growing spatial unit, a city. And the analysis of the way in which a propulsive industry may induce growth in other parts of an economy has become an analysis of how the growth of a place may induce growth in other places. And thus, growth pole doctrine has become an eclectic synthesis of ideas whose introduction into practice soon demonstrated their limited viability. It was not surprising then that few of the designed growth poles around the Mediterranean and inside southern European countries succeeded. And this was not restricted to Europe only. General conditions deteriorated continuously, and this led to a search for 'alternative' development strategies in the 1960s and 1970s.

2.2 EMPLOYMENT INTENSIVE DEVELOPMENT AND DECENTRALIZATION VERSUS NEO-LIBERALISM

During the first half of the 1970s, liberal research concentrated on identifying more 'equitable' policies for the Keynesian state towards employment-intensive development strategies. This new emphasis was initiated by the realization that, through the Keynesian macro-planning of previous decades, no balance between growth incentives and redistribution of growth was reached (Emmerij, 1982). Keynesianism involved a comprehensive strategy and ideology aimed at demand-driven development based upon state planning, directed government investment, media-shaped consumerism and welfare programmes designed to assuage working class pressures and dampen social unrest. Despite the most fanciful optimism, expressed in declarations about cyclical crises that have been officially eliminated, Keynsian state planning began to disintegrate, and a 'new insight' appeared around the notion of the so-called basic needs strategies. Specifically, this means giving priority to the consumption needs of the masses, creating an internal mass-market for basic commodities and changing economic development directions to create direct and immediate benefits in welfare, rather than as a result of a long-term trickle-down process. This new insight aimed at redistribution, with growth as the hub of development policies (Seers, 1969).

While the 'production' of these new development policies was concentrated in western prestigious institutions (in the UK, France, Switzerland and the USA), its 'consumption' was mainly oriented towards the capitalist world's periphery. Interest within the industrialized capitalist countries was limited, despite the efforts made by liberal planners.

These alternative development strategies emphasized decentralization, appropriate technology, energy conservation and ecological considerations and were collectively described as the 'soft and self-help path' (Illich, 1980). They were an amalgamation of liberal egalitarian Utopia; selected experiences from China, Tanzania and Yugoslavia; and critical theoretical comments from within the United States and western Europe. Soon, however, it became evident that planning for basic needs, decentralization and soft-path strategies, while they were unable significantly to distribute growth, under existing capitalist relations, could provide a new frontier for capitalist development in the conquest by planners and the market of the informal sector, which centralized, capital-oriented strategies could never successfully control. This was not only an economic issue (e.g. expansion of consumption and standardization of work outside the factories) but a major political one: it involved controlling on the one hand the mass of unemployed people crowded into big cities or dispersed in rural areas; and on the other providing a framework for absorbing small and medium size productive units working outside the official market.

The new development paradigm of decentralization and soft-path strategies, however, did not attract the enthusiasm of the ruling military élites of the time in Portugal, Spain and

Greece. The same was more or less true for other parts of the world where agropolitan development was proposed as the new substitute for the unsuccessful growth pole doctrine (Friedmann and Douglas, 1975). ILO's 'self-reliance' proposal in 1978 had no important success either. Apart from major economic and structural reasons, there was an important political one: while the new strategy could incorporate the informal sector, it could never work without some, even small, steps towards democratization of public life. Authoritarianism is by definition against decentralization of decision making and popular participation. So it is not surprising that southern European governments introduced the decentralization philosophy in their regional policies at the beginning of the 1980s in a period during which other European governments turned inward to a neo-conservative philosophy.

A notable exception was Italy, where research and particular attention was given to the different reproduction patterns that reflect and influence capital restructuring. Studies of developing and backward regions, both rural and urban, showed the persistence to various degrees of subsistence 'economies' based on activities for the direct consumption of members of the household (Paci, 1979). In addition, research in industrial relations demonstrated the spread of forms of organization of informal work at home (e.g. in the sectors of clothing and salaries below the legal minimum, which enables many companies to compete at international level). The latter received major attention in Italy, with the well known debate around '*fabrica diffusa*' (diffused industrial production) and local labour markets (Magnaghi and Perelli, 1978; Garofoli, 1974; Arcangeli *et al*., 1980; Garofoli, 1983).

In the late 1970s the new 'soft and self-help' strategy was suggesting a paradigm-shift towards a 'more equitable spatial development', via 'selective spatial closure' and 'development from below' (Stöhr and Tödtling, 1977). Significantly in the present context, that case is based on neo-populist notions of development and incorporates the neo-populists' concern for the ill-effects of centralization and large-scale organization (Gore, 1984).

The best current example is to be found in what Soja (1983) describes as the 'New Territorialism', exemplified most clearly by Friedmann and Weaver (1979) and Stöhr and Taylor (1981). For the New Territorialists, what appears to be happening today is presented as a profound shift in ideas about reality, negating the immediate past while repeating similar historical swings in theory and attendant practice. Territorial regional planning, Friedmann and Weaver argue, offers a better alternative approach to the task of promoting regional development. Territorial planning, in this respect, will be an 'endogenous' activity, conducted within the region where its decisions take effect. For Friedmann and Weaver the shaping rhythm of history is the shifting destinies of territorial versus functional power; while for Stöhr it is the changing interaction between small scale - from 'below' - versus large scale - from 'above' - societal action. The interpretation of this new phase includes an emphasis on the big multinationals with their extension of 'functional - big ' as opposed to 'territorial - small' forms of social organization. Friedmann sees them as an 'emerging world historical force' towards functional integration and the destruction of territorial life.

The weaknesses in this framework are rooted in its idealism and the related functionalism which arises from confusing normative assertion with rigorous theoretical formulation. The territorial community is idealized, almost 'fetishized' , into the virtually exclusive repository of cultural traditions, small-scale community action and social value, a conceptualization that resembles the Hegelian assertion of the territorial state as carrier of the Absolute Spirit (Soja, 1983). By ignoring or oversimplifying the role of capital, class struggle and the capitalist state, 'development from below' and new 'territorialism' are likely to be trapped in another cycle of good intentions and unmet expectations, paralleling the experience of the post-war planning theories they criticize.

The latter could be recognized in the spatial policies of the socialist government in Greece in the period 1981-85. The rhetoric about decentralization and people's participation not only was confronted with the hard reality of selfish Greek capitalism but soon fell apart due to contradictory government policies. The example of Athens Greater Area is illustrative. The initial decision to stop Athens' growth was based on typical neo-populist arguments against its size, the parasitic activities it contained, its social and environmental problems and the like, and in support of the rest of rural Greece, which suffers from centralist policies of previous rightist governments. The strategic plan for Athens, designed in 1983, postponed all major public infrastructure, including a new international airport, the extension of the one-line subway system and a programme of highway construction. In addition, location of new industries or extension of existing ones were prohibited, while an odd-even plan for traffic control was introduced in the CBD. The dramatic fall, however, of socialist votes in the Athens area in the 1984 European elections (from 47 per cent to 37.1 per cent) and pressures from powerful contractors' and engineers' lobbies changed these plans. Athens became again *l'enfant gâté* of public spending on infrastructure, while the location of new industries became now necessary in order to 'modernize' the productive structure of the area. The new policies could not be classified as inappropriate for Athens. The problem lies in the decision process, which easily changed from one end to another, always 'from above', showing in the last analysis that neither 'socialist alternative planning' exists nor any enthusiasm for people's participation.

The kind of implementation and procedural problems in spatial policies identified in Greece change their content if we move to the rest of socialist southern Europe. The intellectual currents that was most influential among socialist governing circles in Spain, France, Italy and Portugal have taken up the political terrain previously held by liberal-centrist forces. Surprisingly, however, there was increasingly a detached critical distance from any measures that promote positive state intervention in favour of greater equity, including regional policies. Of greater surprise was the fact that this dissociation from state intervention was not the outcome of self-criticism against the traditional 'statist' view, used so far by the left as an antidote to the free market economy. On the contrary, the new posture was one that emphasizes the importance of 'discipline of the market place', the adaptation of labour to the changing demands of the market and the adaptation of capital to the needs of restructuring. As Petras (1984) characteristically observed, 'this is Reaganomics with a socialist gloss'.

In fact, a reading of the headlines of the newspapers published in OECD countries in the 1980s indicates that virtually all countries are now critical of the problem of rapidly increasing welfare outlays compared to the stagnating income of the state: they have proposed cuts in wages, salaries and public expenditure; and they accept low, zero or even negative, rates of economic growth (Emmerij, 1982).

Under the pressure to reduce public expenditure in all domains, including regional policies, most countries seem to be driven more and more into a defensive and inward-looking corner. The three unsuccessful meetings during 1983 and 1984 of the Prime Ministers of the EEC are a typical example of this. Neo-liberalism wants to decrease the role of the welfare state and to stimulate private initiative and hence the role of the market. We do not need to go as far as 'Reaganomics' or to Thatcher's 're-privatization of Britain' to hear the new key words: adjustment, austerity and self-reliance. For neo-liberalists both at home and abroad these words mean that individual developing units (factories, sectors, regions, countries, etc.) should adjust by their own means, and resources to the existing economic order (national or international), rather than that order adjusts to the needs of individual units by providing financial and institutional help. In addition, outlays for development cooperation (sectoral and regional) should be reduced, and development aid should concentrate on emergency situations only (Mandi, 1983).

In southern Europe, Spanish socialist economic policy involves technocratic planning from above, aimed at allying the state to industry. Unlike the case of France and Greece, in Spain there was no initial burst of reformism, but an explicit rejection of nationalization (Petras, 1984). In Italy, policies point toward promoting the competitiveness of private industry through lowering wage costs. The socialists are being aided by the trade union affiliates of both the Socialist and Christian-Democratic unions as the government's drive to promote productivity and to increase exports is central to the success of its economic programme.

2.3 SOME CONCLUSIONS

All of the approaches in regional development theory that have been discussed are attempts to introduce the spatial dimension in economics. The answers will vary between countries and over time. But it should be evident that there are sharp disagreements within mainstream regional development theory about how the geographical pattern of development may be explained and how a more desirable pattern may be achieved through planning intervention.

The ideas outlined do not summarize the whole spectrum of the field. The account has only focused on a few important contributions in southern European regional policies. This analysis demonstrated three important deficiencies that characterize the majority of the presented theories. First, their concern about 'space' has been reduced to the simple location of social phenomena in geographical space, where cost-distance, time and physical characteristics play the decisive role. From this followed a 'displacement' of the region, as a pre-defined spatio-statistical unit. Regions have been analysed as spatially separated homogeneous units, and then the nature of interregional relationships has been studied. This has been a common procedure in the elaboration of neoclassical, Keynesian and neo-populist approaches, but these 'regions' are in fact merely 'spaceless points' for the location of economic activities. This tradition has been characterized by R. Sack (1974) and C. Gore (1984) as the 'spatial separatist theme'.

Secondly, their view about 'development' was also reduced either to simplistic economic calculations of regional income or to highly Utopian 'partisan social reconstruction from below'. In both cases, capitalist production and distribution, accumulation, exploitation and class struggle remain unnamed or misinterpreted, as if everything produced in each region were the result of an 'Invisible Hand's' operation. The mode of conceptualizing the relationship between society and social change remains unaltered, thus leaving the emphasis of procedure empty of content. In this respect, debates or pseudo-dilemmas about centralization versus dispersal, labour-surplus versus labour-intensive strategies, big versus small, territory versus function, development from above versus development from below, and so on, seem of little relevance.

And thirdly, like many development studies, regional development theory has grown into a major state policy instrument without an adequate analysis of the state itself. The state and political theory remained the _terra incognita_ for regional development theories. When it has been analysed this was only for its administrative functions, something outside and above the process of development. It remains the supreme institution of common interest which guarantees law and order within a given territory. Any discussion about the connection of state's power with class structure, the conflicts between institutional apparatuses and class interests, etc., are totally ignored or misinterpreted, as in the neo-populist approaches.

Thus, the entire body of regional theory rests upon a method of abstraction and a number of false assumptions which, in a contradictory way, are the source for both its major inadequacies and its powerful mode of reproduction. On the one hand, the inadequacies built in such approaches do not guide them to innovative explanations of regional phenomena or to successful regional policies; while on the other, this broader idealistic and empirical framework enables bourgeois regional planners to replace endlessly 'bad' theories by 'good' new ones without dispute about the limitations in their method and assumptions.

Export base and capital/labour mobility theories were replaced by growth pole - growth centre doctrine which itself was superseded by agropolitan development and decentralization strategies; currently we see new best sellers competing for attention: 'selective territorial closure', 'regional self-reliance', 'development from below'. Despite logical improvements and the use of some radical rhetoric (after the destabilizing effect of Marxist criticism), all of these approaches suffer both in their theory and practice from the same inherent limitations described previously. With society always resting on the 'brink of crisis', solutions have become no more than moral incentives resembling the hell-fire and brimstone sermons from the pulpits of another century.

In this very specific sense, the practice of mainstream regional theory is not and cannot be truly analytical or socially critical. This does not mean that it is necessarily wrong or useless. No matter how brilliant, no matter how dissatisfied with modern life, regional planners, as the simple consequence of adhering to the premises of their method, are unable to analyse the origins and causes of regional problems that so much of their liberal protest is dedicated to rectify.

NOTES

[1] In 1957, Gunnar Myrdal, who had acquired

fame as Director of the UN Commission for Europe, published his seminal book, Economic Theory and Underdevelopment, introducing the notions of backwash and spread effects.

[2] This is a basic assumption of 'dualism'; in other words the existence of two separate and independent social and economic structures within the same social formation. A general analysis and cricitism of dualism was made by S. Amin (1972) and for its regional application by D. Slater (1975).

[3] Similar propositions were made by planners in southern Europe. See, for Italy, J. La Palombara (1966), for Greece, B. Ward (1964), for Spain, the World Bank Report (1966).

[4] Export-base theory has played a significant role as an economic doctrine for national and regional development around the Mediterranean. It has been associated in the late 1960s with export substitution theories and remains today a powerful model for rapid integration into the international capitalist market.

[5] Through the Marshall Plan the US Military Service had practically controlled all planning decisions in Portugal and Greece. And, when the Germans offered to Greece a steel mill and a bauxite mining system as part of war re-erections, the US Military Head of State vetoed it, see McNeill (1978).

[6] L. Lefeber was an advisor for regional policies to the Greek government in 1966 and in 1982, when the new Socialist government took office.

Chapter 3
UNEVEN REGIONAL DEVELOPMENT AND MARXIST THEORY

We have, in the previous chapter, noted the various approaches to the problem of uneven regional development, as well as their failures in dealing effectively with the regional crisis. Uneven regional development, however, is an existing phenomenon whose analysis cannot be restricted to proving the ineffectiveness of dominant theories; in a normative sense, one must investigate possible alternatives.

The primary objective of this chapter is to trace briefly the evolution of Marxist theory associated with uneven regional development in the light of the preceding discussion. Two points must be made clear. First, there is no Marxist Regional Theory as such, just as there is no Marxist geography, economics, sociology, etc. We can only speak of a Marxian approach to all of the above if we are to avoid reproducing bourgeois epistemological divisions. When Marx said that he was not a Marxist, we should take him seriously and not shrug this aside as a *bon mot* (Wallerstein, 1983).

Secondly, Marxist studies of uneven regional development, while they have made several advances in the field, are also characterized by several limitations. Much work which carefully criticizes traditional development theory shares many of its faults when alternatives are proposed. As Massey (1977) argues in relation to industrial location studies, Marxist approaches often tend to over-emphasize the importance of economics and structural determination, producing in this sense reductionist and monocausal explanations.

As a partial effort to overcome these problems,

this chapter is devoted to a consideration of the thesis that there are no iron laws of motion of capitalist development, nor a pre-defined theory capable of explaining socio-spatial unevenness. What we are looking for is an alternative framework that is both sufficiently comprehensive to explain the multidimensional face of uneven regional development and sufficiently simple to provide a broad set of guidelines that can in turn be made more specific in particular historical contexts (as in southern Europe, for instance) and serve as a basis for policy formulation and political action.

3.1 BEYOND SOME THEORETICAL OBSTACLES

It has become common practice among geographers, sociologists, regional planners and others working with a Marxian perspective to argue that, within the classic work of Marx, Engels and Lenin, there existed a strong geographical emphasis but that this emphasis and orientation remained weakly developed by successive generations. Similar observations have been made by political scientists, in terms of the political element, the transition from capitalism to socialism and the theory of the state; and by sociologists in terms of social classes, ideology and culture. Relatively little attention, however, has been given to explaining _why_ spatial analysis remained so weakly developed for so long a time.[1] Unravelling the history of this submergence of Marxian spatial analysis is a major project still to be undertaken. It is possible to offer meanwhile a few promising areas of inquiry and explanation.

a. Marx's failure to complete _Kapital_ with volumes dealing with world trade, colonialism and the geographical expansion of capitalism, only hinting at their possible context in the late-appearing _Grundrisse_.[2] While Volume III of _Kapital_ presents a concretization of Marx's theory, especially in the labour theory of value, an analysis of issues relevant to regional development, such as capitalist markets (national and international), the articulation of pre-capitalist and capitalist modes of production, the role of the state - in essence the foundations of an analysis of 'uneven development' between sectors, regions and nations was lacking.[3] In the absence of these sources, heavy emphasis was placed upon the largely aspatial closed-system theorizing of *Kapital*'s first volumes. Although Marx never fails to illustrate his arguments with specific historical and geographical examples, Volumes I and II remain encased in the simplifying assumption of a closed national economy, an essentially spaceless capitalism systematically structured almost as if it existed on the head of a pin.

b. A long established anti-spatial tradition in western Marxism originating in the restraining influence of the Second International and later of Stalinist dogmatism. The interwoven roles and central significance of spatialiity, politics and the state were submerged by adherents of Second International in a sterile economic reductionism, a positivistic Marxian 'scientism', emphasizing technocratic thought. This attitude created two orders of reality. The first, that of necessary laws of evolution, was theorized in a rigorous fashion to present as history the causal relation between base and superstructure. The second, an order of facts, amounted to no more than the sum of empirical circumstances, given the absence of any perspective which would allow a less deterministic analysis of the conjuncture, the particular historical/geographical moment (Laclau and Mouffe, 1981). For example, the specificity of the peasantry could safely be ignored, since it was a sector destined to disappear in the course of capitalist development. Furthermore, ideology and politics did not constitute reality, since they were simply expressions, representations of a process inscribed in production. Space, territory and regional consciousness were similarly off the theoretical agenda (along with gender, every day life, the environment, etc.), since spatiality was considered only as a physical container, a neutral terrain for the productive forces determined by the laws of capitalist accumulation.

Lenin himself was not responsible for this reductionist interpretation of politics and the state. He assigned an important position to the political in his notion of a 'specific critical conjuncture', dominated by a displacement in the relation between classes, a situation with potentials for revolutionary change. But, under Stalinism, the new political line (still embraced by certain Communist Parties) installed a permanent dualism between the political logic of Leninism and the economic logic of the Second and Third International. The primacy of the political was to be preserved for critical moments only, while economism continued to dominate at all other times (Poulantzas, 1978).

c. Changing material conditions in the development of capitalism and their reflection in conditions of accumulation and exploitation. H. Lefebvre has argued in the 1970s (for example, in La Pensée Marxiste et la Ville) that, during the nineteenth century and into the early twentieth, the production of geographical space was accommodative, conformant and directly shaped by the market. In addition, given the massive urbanization associated with expanding industrialization, the reproduction of the labour force was much less crucial an issue than the process of direct exploitation through a system of subsistence wages and the domination of capital over labour at the point of production. In the extraction of absolute surplus value, and in the simple reproduction process, time was more important than geographical space.

In the post-war period and until the 1980s, the conditions which underlie the continuing survival of the system have changed. Capitalism, on the one hand, has been forced to shift greater and greater emphasis to the extraction of relative surplus value through technological change and modifications in the organic composition of capital, and it is increasingly dependent on the reproduction of the labour force and the general social order. On the other hand, the growing complexity of contradictions in the system made state 'positive' intervention a generalized principle. These developments required the construction and control of social systems to secure the smooth reproduction of the social relations of production, which has become the key issue in welfare capitalism.

The hard reality of neo-liberalism since the 1980s in many European countries and the restricted changes introduced by southern European socialist governments presented an opposite commitment: a massive decline of state intervention in the spheres of circulation and production. Even in those cases, however, the state still is playing a huge role in securing the conditions for the reproduction of capitalist social relations. The abandonment of whole regions 'to their fate', the loss of thousands of jobs, and the restructuring of capital - to mention only few well known examples - are the new style of 'negative' state intervention. The aim remains however the same as in the past: making capitalism function profitably. In this process, uneven regional development and the production of regional space play an increasingly important role, and under such conditions an intensified interest in Marxist theory and practice is expected.

Thus, regional space cannot be considered as a passive container of productive forces, but is an active element in the complex interplay of social relations of production. This dialectical role of space was emphasized by N. Poulantzas (1978) in his last major book:

> ... I am not speaking of some mechanical causality according to which pre-existing relations of production give rise at a subsequent stage to spatial ... matrices. Themselves implied by the relations of production and social division of labour, these matrices appear at the same time as their presupposition - in the sense that Marx gave to the term logical priority (Voraussetzung) as opposed to 'historical preconditions' (historische Bedingungen). (Poulantzas, 1978, p.99) (emphasis in the original)

The relation between Poulantzas' 'spatial matrices' and the social division of labour is not a contradiction-free process. In fact, the development of systematic Marxist analysis of uneven regional development coincided with the intensification of social and spatial contradictions in both core and peripheral countries in the late 1970s and early 1980s.[4]

3.2 PRECURSOR: A. GRAMSCI

A major contribution to the development of Marxian analysis related indirectly but constructively to uneven regional development can be found in the work of Antonio Gramsci, whose concepts are now experiencing a widespread revival among Marxists of different specialized orientations (Buci-Glucksmann, 1975; Mouffe, 1979; Sassoon, 1980). In part, Gramsci's work is useful in spatial analysis because it contains some well-elaborated statements on urban and regional problems in Italy during the 1920s (e.g. the city-country contradiction, regional backwardness in the Mezzogiorno, urban development in Turin, housing problems in other major Italian cities, alliances between urban and rural proletariat). But, of even greater relevance to contemporary developments in Marxian spatial analysis was Gramsci's effort (a) to re-establish the Marxist dialectic (against the vulgar materialism of the Second International and the reformism of social democracy); (b) to focus attention upon civil society through the analysis of political, ideological and cultural problems in capitalism (against the prevailing economism of the time); and (c) to analyse the functioning of the capitalist state.

In his emphasis on the 'ensemble of relations', Gramsci criticized what he called 'absolute historicism' and clearly differentiated both material and social conditions of production within what today would be called the social formation. The specificity of the social formation (the articulation between various modes of production, one of which is dominant) represented the mode of production concretized in time and space, history and geography. For Gramsci, this was the indispensable 'conjunctural' framework upon which revolutionary strategy had to rely. A spatial problematic was not explicitly raised as such, but its foundations were clearly evident in the spatial relations of the social formation and the particularities of place.

Revolutionary strategy for Gramsci was aimed at three interrelated areas, all of which are relevant to the regional problematic. The first involves his emphasis on the political and cultural/ideological structures in the social formation embodied best in his concept of hegemony.[5] By hegemony Gramsci meant a whole body of practices and expectations, penetrating every day life in a society - including structures and activities such as trade unions, the educational system, state administration, the church, the language, the distinction between city and countryside, the backwardness of certain regions, etc. - which are in one way or another supportive of the established order and the class interests that dominate it. Hegemony of the ruling class over a social formation in this sense is diffused by agencies of ideological control and socialization and to the extent that this has been internalized by the broad masses it becomes part of 'common sense' (see also, Hoare and Novell-Smith, 1971; Buci-Gluksmann, 1975; Boggs, 1976). When the masses, however, become able to develop their own counter-hegemony, i.e. a set of practices and beliefs that questions the ruling class hegemony, periods of upheaval and crisis arise. As Gramsci himself argues:

> If the ruling class has lost its consensus, i.e. is no longer 'leading' but only 'dominant', exercising coercive force alone, this means precisely that the great masses have become detached from their traditional ideologies and no longer believe what they used to believe previously. (A. Gramsci, Prison Notebooks, p.276)

This first area of emphasis relates directly to the second: the role of exploitation of the working class at their place of residence, the point of consumption and reproduction *vis-a-vis* the point of production *per se*. This not only reopened the 'housing question' to new considerations but was also in conflict with both the economism and productivism of the Second International and the 'unionism' of the Italian Socialist and Communist parties of his time. In addition, Gramsci's stress on hegemony and his work on popular culture and control over every day life, indicate the important role of organizational efforts by popular forces at the local level (e.g. local councils).

Finally, Gramsci's emphasis on the *revolutionary historical bloc*, his notion of alliances in the popular movement fighting for the same common goals, pointed out the need to understand the particular conjunctural crises of capitalism; that they were not only economic but political, cultural and ideological; and that they involved not only production but reproduction processes as well. In *Prison Notebooks*, Gramsci saw the growing complexity of modern capitalist society raising political, cultural and ideological struggles to a new level, as the state relied increasingly upon hegemony and security of reproduction of the dominant social relations rather than on force and oppression. Revolutionary consciousness thus came to be rooted in 'the phenomenology of every day life'.

The relevance of Gramsci's work for spatial analysis with a Marxist focus has been acknowledged by others as well (Dulong, 1978; Nocifora, 1978;

Soja and Hadjimichalis, 1979; Hastaoglou, 1982; Cooke, 1983). Dulong (1978) points to the growing interest in Gramsci in France since May 1968 and the major importance of his analysis on the relationship between state and civil society. Through Gramsci, Dulong argues, a critique by Poulantzas and others became possible against some versions of Marxist sociology that was dominated until then by functionalism and economism. Hastaoglou (1982) and Cooke (1983) provide a more elaborated synthesis of Gramsci's ideas in a spatial setting, centered around the concepts of civil society, the city-country contradiction, hegemony and the specificity of locality and territory. Gramsci's analysis of Italian history, based on the city-country contradiction, is not just a metaphorical use of space and territory; in Italy - as well as in other southern European countries - social relations and class conflicts revolve around a central territorial component. But such a contribution is to be further developed. As Nocifora (1978) argues, the work of Gramsci in Italy, which could have provided a solid basis for investigating the social and spatial relations of production, has instead been used for the most part as a convenient excuse for not having to venture into those areas. And Nocifora concludes:

> The regulation phrases about the unity of the struggle between workers of the north and the peasants of the south were repeated *ad infinitum*, but they were rarely translated into concrete political practice. (p. 363)

Whether one agrees or not with this, it is nevertheless true that the major interest in Gramsci, not only in Italy but in the rest of Europe as well, remains on politics and strategies rather than specific aspects of everyday life (Salvatori, 1979). As L. Paggi (1979) argues, Gramsci represents both a continuity and a rupture with Leninism. While he extends Lenin's major argument that history does not proceed according to a simple functional logic and that every articulation and radical change depend on a political intervention by the revolutionary subject, he does not confine the primacy of the political to revolutionary conjunctures alone, but makes it the *articulatory principle* of every social situation.

It is precisely this Gramscian notion of the political as the articulatory principle that lies at the heart of the present problem of uneven regional development in most European countries and the production of space in general.

3.3 CURRENT CONTRIBUTIONS AND DEBATES: AUTONOMOUS VERSUS TRANSFER OF SURPLUS DEVELOPMENT THEORIES

Current Marxist research and debate on the regional question has had four main sources of influence: (a) urban sociology and geography, influenced by Marxist methodology; (b) underdevelopment theories; (c) theories of social classes and the capitalist state, and (d) capitalist crisis theories. It is not my intention here even to attempt a detailed review of the massive empirical and theoretical work concerning these issues. There already exists a number of good reviews connecting these sources with the question of uneven regional development (Slater, 1975; Goldsmith, 1978; Harloe, 1978; Massey, 1978a; Lebas, 1978; Soja, 1980; Cooke, 1983). The discussion which follows is therefore brief, selective and only indicative of larger trends and emphases.

During the 1960s and 1970s Marxist research on the regional unevenness of capitalist development succeeded in establishing itself as an alternative analytical framework to neoclassical and liberal theories. Initially built upon general laws and abstract formulations, upon approaches 'borrowed' from international development theories and theories of capital accumulation, geographical analysis using Marxist tools and methodology was and remains highly heterogeneous (Massey, 1978a; Soja, 1983). At the risk of over-simplifying, I will identify two dominant lines of thought within this heterogeneous intellectual tradition (Hadjimichalis, 1984).

The first, based on the process of capitalist production, addressed the self-expansion of capital and has put forward the <u>autonomous</u> or <u>semi-autonomous development thesis</u>. In this case, different geographical areas (however defined) grow or decline depending on concretely different production processes, types of product manufactured by firms located in these areas and the diferent use-value circumstances of capital and labour operating in these places (Walker, 1978). Development is dependent upon putting together the necessary conditions for profitable accumulation (mainly via exports). Sectoral composition and dynamics are major causal forces shaping regional differentiation (Markusen, 1983). This line of thought originates from Lenin and Bukharin, who centred their analyses of uneven development on the contradictions of accumulation that exist _within_ certain areas (Lenin, 1973; Bukharin, 1973). In their view, it is possible for various capitalist sectors to grow relatively autonomously from other sectors: this offers local investment opportunities but also leads to the accumulation of enormous masses of capital that drive down the rate of profit. As Lenin observed:

> ... surplus capital then seeks 'young areas' external to the capitalist sphere: in these areas profits are usually high, for capital is scarce, the price of land is relatively low, wages are low, raw materials are cheap. (Lenin, 1973, pp.73-7)

This association of different rates of profit with different areas is also analysed by Harvey (1975), who uses Marx's concept of the 'annihilation of space by time' to emphasize capital's goal to lower its time of circulation and speed its self-expansion.[6] In an attempt to use the profit cycle model, Markusen (1983) argues that super profits in certain firms are a return to capitalists due to lower production cost (socially necessary labour time) or to the shorter time that a household spends in reproducing labour power. Following the same argument, Läpple and van Hoogstraten (1980) suggest that the 'fundamental causes of unevenness are the continuous expansion of surplus value production and the increase of productivity, occurring within the same firm at the same place. Without giving much attention to uneven geographical development, Brenner (1977) defended the autonomous development thesis, arguing that neither development in the core nor underdevelopment in the periphery was determined by surplus transfer. Following Brenner and working in geographical setting, Walker (1978) argues that uneven geographical development is a consequence of capital's internal logic working itself out in space due to particular 'use-value arrangements of places'. And Webber (1982) concludes that the regional question concerns the spatially uneven development of branches of production, while agglomeration economies provide the immediate cause of such uneven development.

An interesting common future in almost all the attempts to use the autonomous development thesis in explaining uneven development over space is the conclusion that 'space-specific' or 'locality-specific' characteristics in the spheres of production, circulation and exchange are unimportant. And any attempt to reintroduce socially produced space in modern development thinking tends to either simplistic explanations or to dangerous 'fetishisms'.[7] A good example of this approach is Browett (1984), who concludes that any attempt to introduce geographical factors in analysing uneven development is insignificant. The processes underlying the growth and regional movement of capital, according to Browett, are regulated by the inner law of the capitalist mode of production, while spatial form is a simple consequence of economic factors.

Four serious weaknesses are identifiable in the autonomous approach to uneven geographical development. First, although it focuses on accumulation, it submerges the importance of circulation and exchange. Production in itself is the driving force for development, while anything else is subordinated to it. Secondly, and unsurprisingly, regional development is conceived as a matter of the autonomous growth of local productive forces only, thus denying any significant relationship between sectors and firms in different geographical units in the outside world. Thirdly, because the main focus of abstraction is upon the logic of production and accumulation, there is little reference to the role of class mobilization, political action and the role of the state. And fourthly, there is a tendency to regard everything that occurs in space and time as conducted on behalf of capital, ending either in cyclical arguments or in economic reductionism. Thus, the spatial organization of society is taken as a simple by-product of the capitalist production, as a passive container of productive forces.

A second approach places major emphasis upon the sphere of capital circulation and exchange, ignoring (in some cases) production relations. Through circulation and exchange, surplus produced by workers in some areas ('the periphery'), is extracted and transferred (via a number of mechanisms) to benefit capitalist development in other sectors and areas ('the core'). This alternative we call <u>transfer of surplus development thesis</u> (Emmanuel, 1969; Baran and Sweezy, 1970; Lipietz, 1977; Amin, 1977; Liossatos, 1979). Development is said to be dependent on the economic and political strength of the core's social classes, which are able to accumulate wealth and power while the periphery is 'drained' via surplus-transfer. Otto Bauer (1924) was a direct supporter of the surplus-transfer thesis, arguing (against Lenin and Bukharin) about the appropriation of surplus-value from less developed areas, when these areas had 'less capital' compared to developed areas.[8] Four years later, in 1928, O. Kuusinen, at the 6th World Congress of the Comintern, defended a thesis that capitalist penetration in the periphery of the world system under the aegis of imperialism was not only a source of surplus extraction to the benefit of the centre but it also created a 'bottleneck' to industrialization and resulted in stagnation. These ideas acquired, after World War II, wider theoretical legitimacy through the 'development-of-underdevelopment", 'dependency' and 'core-periphery', schools of thought (Frank, 1979).

Dependency theorists argue that the process of capital export via multinationals permits the capture of profit rate in the periphery, which is higher. The surplus thus captured is repatriated: the development of the forces of production in the periphery is blocked, while this surplus enriches the centre, where it contributes to a further advance in the development of productive forces. Such views borrowed from international relations have been transplanted to the regional level with obvious problems in terms of empirical differences as well as theoretical ones, as correctly pointed out by Slater (1975) and Massey (1978a). In this respect, the dependency school came close to liberal and neo-populist views of polarization (Gore, 1984). Finally, an elaborated version of the transfer of surplus development thesis is the world-system theory, the main proponent of which is Wallerstein (1974, 1976, 1983). Wallerstein introduces a clear political basis for a dialectical approach to social and spatial phenomena, arguing that:

> The operation of the (capitalist world) system, once established, revolved around two basic dichotomies. One was the dichotomy of class, bourgeois versus proletarian ... The other basic dichotomy was the spatial hierarchy of economic specialization core versus periphery ... The genius, if you will, of the capitalist system is the interweaving of these two channels of exploitation. (Wallerstein, 1976, pp.350-1)

Without accepting some of the functionalist

argument of other dependency theorists, Wallerstein introduces directly the role of the state in describing the transfer of part of surplus from periphery to the core, what he calls 'the particular space-time nexuses', which are 'hidden' in historical capitalism (Wallerstein, 1983, pp. 29 and 31).

The surplus-transfer concept, however, gained its major theoretical status with the Greek Marxist economist, Arghiri Emmanuel (1969) in his book Unequal Exchange and the long debate that followed it. Emmanuel distinguished three forms of unequal exchange: first in the 'broad sense', when only differences in the organic composition of capital exist; second, in the 'narrow sense' with differences only in wages; and third, in the 'strict sense' with differences both in organic composition and wages. Interregional polarization develops as extroverted versus introverted accumulation is taking place, or as regional differences in both the organic composition of capital and labour costs increase.

The main problem with these accounts of uneven geographical development is their incomplete abstraction of the determinants of the processes to be explained. First, there is an over-emphasis in the sphere of circulation and exchange, in contrast to the autonomous development thesis. Secondly, their prediction of a continuous 'underdevelopment/ stagnation' of the periphery through surplus transfer was over-simplified, as international (e.g. Brazil, South Korea, Taiwan) and regional (e.g. the role reversal of regions in the UK, USA, France) experience has shown. Thirdly, they use an unclear definition of exploitation via external factors and exchange, thus ignoring local social relations and class struggle. This rests upon a mainly technological explanation of the sources of increased productivity and, hence, wage increases. Fourthly, they assume equilibrium in the capital market (internationally and nationally) which implies that capital movements (the more mobile factor) have already resulted in a world/ regional equalization of the rate of profit. And fifthly, some of the authors - with the exception of Mandel and Wallerstein - following a functionalist categorization between core and periphery failed to identify social classes within particular areas, thus arguing indirectly about place exploiting place, confusing between space as a separate structure and space as a social product.

It seems, therefore, that, while considerable progress has occurred in the theorization of the process of uneven geographical development - due to deeper understanding of the underlying processes associated with capital accumulation and circulation - such progress may be undermined by an emerging dogmatism in both dominant lines of thought. Struggling to be serious and rigorous in their application of Marxist methods, many (on both sides) began to establish certain boundaries beyond which a leftist analysis could not reach. The process seems to be analogous with what Soja (1980) has described in his analysis of the socio-spatial dialectic:

> ... instead of sensitively probing the mix of opposition, unity and contradiction which defines the socio-spatial dialectic, attention has too often been drawn to the empty question of which causes which or to endless arguments about pre-eminence. (Soja, 1980, p.208)

Instead of focusing upon a fundamental linear idea of capital accumulation - either autonomously or via surplus transfer - I propose a third view which focuses upon both production and exchange and upon local social classes and the state. Theories which treat exchange as the dominant source of exploitation and the principal factor of development in regions have been found inadequate. However, it has also been seen that a focus on production requirements and accumulation imperatives, to the exclusion of exchange relations, tends to be closely associated with an equally inappropriate economic reductionism (Cooke, 1983).

What must be clearly demonstrated, then, is how and in what ways production, circulation and exchange relations connect accumulation to specific localities, and whether a relation could be established among these localities based on the labour theory of value. Before I pass to this level of analysis a short comment on the labour theory of value and the unity of production and circulation is necessary.

3.4 THE LABOUR THEORY OF VALUE (LTV) AND THE ARTICULATION OF PRODUCTION AND CIRCULATION

All class societies, whether capitalist or not, can be understood as consisting of two broad categories of people: direct producers, the women and men who produce the goods and services which allow society to continue; and non-producers, those who live off the production of others. In this respect there exist certain non-economic social relations between producers and non-producers, which first secure the existence of the non-producing class (e.g. through institutions and ideological legitimization or direct oppression) and secondly provide to the non-producers the social mechanisms necessary for the appropriation of the social product. Thus, capital accumulation must be understood as the reproduction of capitalist relations of production, and capitalist development must be understood as this reproduction on an ever-expanding scale through the conversion of surplus value into new constant and variable capital (see also, Wright, 1979).

Value is the unique characteristic which only human labour can give to products and commodities by transforming them. A product has value only because human labour in the abstract is embodied in it. Exchange value is the essential property of commodities that differentiates them from the simple products of labour in general, i.e. from things not produced specifically in order to be exchanged. A commodity is a use-value for its non-owner and a non-use value for its owner. Hence the need for exchange. In addition, value provides us with a useful concept to understand the social mechanism of capitalist production, the antagonistic relation between capitalists and workers. If, for example, one wants to analyse the mechanisms by which the social product is divided among classes, it is necessary to have units for measuring different quantities of products. The most obvious metric, of course, is simply the price of commodities. In fact, this has been in practice the solution to the problem for most bourgeois economists. But, as Emmanuel (1969) argues, using price as the metric for comparing quantities of products raises the question: what is it that the money-price attached to a commodity is measuring? What is the theoretical content of the quantitative dimension of the social product that prices tap? If the answer is not to be totally circular, some sort of value theory, a theory of what constitutes the quantitative dimensions of commodities, is necessary.[9]

In this respect, value is not price, and when we talk of the value of a commodity we do not mean its price. Nor is there any implication that the rate of exchange of commodities in the market is strictly proportional to their values. On the contrary, as Marx argues:

> The possibility ... of quantitative incongruity between price and magnitude of value, or the deviation of the former from the latter, is inherent in the price form itself. (Kapital, Vol.1, p.75)

The debate on whether Marx's Labour Theory of Value (LTV) is necessary in the analysis of capitalism and whether it is reconcilable with the actual relations involved, is an old one. It has acquired great popularity through Sraffa's work (1960) and more recently with Steedman (1977) and the neo-Ricardian school. Initially confined to a relatively narrow grouping of economists, the controversy about value theory has spread to wider circles of the left. It is beyond the scope of this chapter to provide a detailed review of the ongoing debate. Instead I will highlight some of the points relevant for the discussion of the geographical transfer of value.

Theories using the Sraffian perspective argue that profits can be considered as a direct consequence on only two factors:; the socio-technical conditions of production and the real wage paid to workers. Other factors of profit have an effect only by virtue of their influence on these two factors. This argument is based on a mathematical analysis of the necessary conditions for formally calculating profits from a set of initial conditions. The categories of LTV, neo-Ricardians argue, do not enter into this calculation at all (Steedman, 1977; Hodgson, 1981).

The critical point of this model, as Wright (1979) argues, is not only to reduce the theory of profits into a simple two-factor account, but rather to argue that other causes have their effect on profits via real wages and technical conditions. Thus, class struggle, political conditions, the state, ideological hegemony and - as I am going to argue later - space, do not have direct effects on aggregate profits. Following Wright's criticism, Shaikh (1980, 1981) emphasizes the fact that, if one's interest in studying profits is limited to calculating profits, then the neo-Ricardian level of abstraction is as far as one needs to go. This is evident, for example in Scott's (1980) otherwise very interesting attempt to use neo-Ricardian concepts in studying urban land. Scott follows Steedman in treating real wages as exogenously determined in order for the economy to yield determinate prices and profits. But he fails to recognize that problems such as capital polarization in space, the spatial differentiation of the rate of profit, and changes in urban land (location and distribution patterns) are not restricted to technical changes only.

If, on the other hand, one's interest is in understanding the social determinants (or structural limitations, according to Wright [1979]) of profits, one needs to move to a higher level of abstraction. This is precisely what the Marxist model of profit determination attempts, using the LTV. As Lipietz (1981) argues, the concept of value, including the magnitude of value, illuminates the whole qualitative and quantitative analysis of price relations, uncovering relationships where the neo-Ricardians see merely discrepancies.

An important problem, which is directly related to my discussion of GTV, is the failure of neo-Ricardians to make the distinction between value and realized value. This rests on their inability to recognize the difference between abstract labour and socially necessary labour. Marx, on the other hand, is explicit about this difference in *Kapital* (Volumes I and III), where he makes a clear distinction between use-values produced for direct use and converted into commodities only when exchanged, and use-values produced for exchange and hence produced as commodities (Shaikh, 1981). Indeed, exchange is a process in which the different labour-times involved in producing these use-values actually confront each other and are eventually articulated into a social and spatial division of labour through the medium of money prices. But can we say that the mass of surplus value gets bigger or smaller depending only on relative money prices? Shaikh (1981) answers as follows:

> ... consider a crisis in which so little of the social product is sold that profit is actually negative (this is a recurrent real phenomenon in capitalism). Are we then to say that, even though workers were exploited and a surplus product produced, surplus-value is itself negative? If we are not allowed recourse to the distinction between value produced and value realised, then of course surplus-value is no longer connected to any rate of exploitation at all. It is merely an epiphenomenon of

> circulation. And so what begins as a tactical capitulation to the neo-Ricardians turns into a rout. (Shaikh, 1981, p.300)

Shaikh's interpretation points directly to the articulation of production and circulation and could provide a way out from both autonomous and transfer of surplus development thesis. Such an articulation derives from the fact that, after the act of production, surplus value is embodied in the value of commodities that must circulate (i.e. be sold) in order for surplus value to be realized as money and for capital to grow. In this respect, the different labours undertaken at different locations are brought into a relationship with each other through acts of exchange. Spatial integration - the linking of commodity production in different locations through exchange - is necessary if value is to become the social form of abstract labour (Harvey, 1982).

A key aspect of the market organization of production and circulation has been therefore its particular spatial configuration. An initial trend towards the spatial concentration of industrial activity had its obverse in the spatial expansion of a decentralized commodity market. While firms clustered at the side of cheap labour or resources, commodity exchange was extended through the economic and political integration of world and national markets. Inter-firm agglomeration economies often enhanced surplus-accumulation potentials. Through these economic and political tendencies, a market-oriented competitive spatial system was conceived with its own dynamic for growth or decline.

In this competitive context, the necessary relation between production and circulation is also contradictory: circulation is the negation of production, and production is the negation of circulation (Lipietz,1983a). The essence of economic crises is to a large extent found in this contradictory relationship. The class nature of capitalism implies the existence of interclass and intraclass conflicts regarding the production and distribution of economic surplus and, in particular, the determination of profit among capitalists versus their costs. Moved by the profit motive and class conflicts, capital is thus driven to expand in the sphere of production and simultaneously encounters a barrier in the sphere of circulation (Harvey, 1982). It is precisely at this point that uneven regional development and, more generally, the organization of space enter into the problematic of capitalist development. Spatial integration is actively produced rather than passively received as a concession to 'nature' or 'history'. And this production of spatial configurations is made up through direct and indirect intervention by private social agents (mainly individual capitalists) and/or collective social agents (the state, local authorities, labour organizations, social movements, etc.). The outcome is that the development of capitalism is beset by counterposed and contradictory geographical tendencies. On the one hand, spatial barriers and regional/local distinctions must be overcome. Yet the means to achieve that end entails the production of new geographical differentiations which form the spatial barriers to be overcome. The geographical organization of capitalism '... internalizes the contradictions within the value form' (Harvey, 1982, p.417).

In summary, the problematic of uneven regional development as has been discussed so far rests on three major assumptions. First, neither autonomous nor surplus transfer theories of development can adequately grasp the complexity of unevenness over space and an alternative view must be adopted. Secondly, an attention and focus on the articulation of production and circulation (derived from the labour theory of value) is necessary, so that a connection between the nature of decentralized commodity production, market exchange and the necessities of sustaining accumulation with different localities and regions could be established. And thirdly, the historical and spatial specificity of these issues must be clearly demonstrated, avoiding deterministic and monocausal explanations. The work of Gramsci, more than any other, is of particular value here, with its specific focus on politics, the state and ideological/cultural factors.

NOTES

[1] In this section I draw heavily from Soja and Hadjimichalis (1979) and Hadjimichalis (1984). For geography and spatial analysis, see the early works of Lefebvre (1968, 1974); Lacoste (1972); Harvey (1973); Soja (1980); and the more recent contributions made by Harvey (1982); Cooke (1983): Gore (1984); Clark and Dear (1984); and others.

[2] Grundrisse appeared first in the Soviet Union in 1939-41 and in English in 1973.

[3] Lenin further develops Marx's argument that capitalism could never achieve equal development in all spheres of production and in all regions:

> ... both uneven development and semi-starvation level of existence of the masses are fundamental and inevitable conditions and constitute the premises of this mode of production (of capitalism) (Lenin, 1963, p.718).

[4] See the points made by Lipietz (1977), Dulong (1978) and Urry (1981b).

[5] In developing his concept of hegemony, Gramsci never denies the primacy of the mode of production in shaping history, but he criticizes the economic determinism derived from the Second International. For Gramsci a true revolutionary theory must take into account the rich interplay of all forces that shape a 'conjuncture' in a social formation; see also, Buci-Glucksmann (1975) and Poulantzas (1978).

[6] Harvey does not stand totally within this tradition, see for example Harvey (1982) and especially chapters 12 and 13.

[7] Most of the scholars agreeing with this general approach tend to underestimate spatial issues in the defence of a pure Marxist analysis against 'spatial fetishism'. Markusen (1979b) in an older essay argues in a classical aspatial economic framework that: '... capitalist production relations in their fundamental logic are spaceless ... there is nothing in the logic of capitalist accumulation that requires spatial differentiation' (p.38).

[8] Otto Bauer (1924) pointed out that:

> Since in the more highly developed area there is more capital to the same amount of labor, this area appropriates a larger share of the surplus value than would correspond to the amount of labor it has contributed. It all happens as though the surplus value produced in the two areas were first of all cast into a heap and then shared out among the capitalists according to each one's holding of capital. Thus, the capitalists of the more highly developed areas not only exploit their own workers but also appropriate some of the

surplus value produced in the less highly developed areas ... (O. Bauer, Die Nationalitätenfragen und die Sozialdemokratie, quoted by A. Emmanuel, 1969, p.175).

[9] This was the main theme of what is known as the 'Cambridge capital controversy', which was concerned with whether or not prices can be considered an acceptable metric of physical capital. For an historical review of the controversy and a Marxist epilogue to it, see Meek (1956).

Chapter 4
THE CONTRADICTION OF EQUALIZATION/ DIFFERENTIATION AND THE GEOGRAPHICAL TRANSFER OF VALUE

It has been acknowledged that the reproduction of the capitalist mode of production is not limited to the simple material reproduction of the means of production as such (e.g. technical reproduction of machines, raw materials and land, and biological reproduction of labour force). Above all, capitalism is dependent on the reproduction of its typical pattern of social relations, i.e. the domination of capital over labour through a complicated interplay of forces. Social reproduction, however, does not occur without producing changes in the character of the dominant social relations. This excludes the idea of an automatic and historically uniform reproductive process internal to the CMP. Social reproduction is rooted in <u>tendencies</u> and <u>contradictions</u>, the dynamic processes which reproduce and at the same time change capitalist social relations.

The purpose of this chapter is to go beyond a surface description of regional phenomena, trying to see what are the <u>underlying tendencies</u> that produce and reproduce uneven regional development. The discussion will evolve around two key propositions:

a. Capitalism survives historically by not fully developing its inherent tendency towards equalization of a number of conditions defined principally, but not exclusively, in the sphere of production and circulation. Economic relations do indeed have the determinant role in a social formation; but not as a result of some internal laws of motion. They are the result of the political and cultural/ ideological intervention of a political agent. It is in this history of survival that uneven regional development arises, based on a principal contradictory relationship, that of equalization/differentiation, where space plays a very important role.

b. The means, the underlying tendencies, by which uneven regional development is maintained over time are (i) differentially localized accumulation (DLA), and (ii) the geographical transfer of value (GTV). Both are sustained and mystified by the state and the local state and its apparatuses and by the wider ideological hegemony of the bourgeoisie.

4.1 THE CONTRADICTION OF EQUALIZATION/ DIFFERENTIATION IN REGIONAL DEVELOPMENT

The approach which is suggested here begins from the historically dominant processes of production and circulation in a social formation and defines uneven regional development in terms of the geographical distribution of the conditions of accumulation in relation to these processes. Accumulation presupposes production and exchange ending in profits, because profit maximization is its very rationale. The competitive search for profits and increasing accumulation in turn involve the constant reorganization sectorally and geographically of the production and exchange process. This is the fundamental basis for uneven development.

Accumulation, however, is not a purely economic process, as it has often been analysed. For the system to function and the reproduction of capitalist society to take place, political and ideological struggles play important roles. The nature of this political and ideological struggle cannot be reduced to determination by the economy (Urry, 1981a). The main merit of Gramsci's approach is to have attempted to theorize the class struggle within capitalist societies, not only around economic issues. More specifically he maintains that the dominant class must be able to establish its hegemony and to form an historic bloc with other major classes. Through hegemony, there is a continuous process of forming and superseding of the interest of the dominant class, both with potential allies and with subordinate classes. Accumulation, therefore, must be considered as part of their wider social framework, stemming from the various forms of class struggle and in turn affecting the forms that such a struggle may take.

During the process of accumulation and competition for higher profits, all capitalists do not succeed equally; the cost of failure is also unequally distributed among different sectors, regions and localities. Different sectors and different firms within the same sector, located in the same or separated localities, have different rates of profit. These differences arise from certain general conditions of capitalist development which are unevenly distributed across sectors and regions and which can be distinguished in three major categories: conditions which affect the sphere of production, i.e. the production/extraction of surplus value; conditions which affect the sphere of circulation/exchange, i.e. the realization of surplus value; and conditions which affect the sphere of reproduction and in particular the reproduction of the labour force. These conditions are part of the 'total social capital' potentially available for each capitalist. They are produced and reproduced by three different types of social practices: by individual entrepreneurial practices, by the practices of the state and local authorities and by the practices of civil society.

Looking closely at these conditions and starting from the conditions that affect the sphere of production, capitalists tend to follow a similar logic which includes actions such as reducing labour costs, increasing productivity, developing technical innovations, intensifying the labour process, increasing scale and - if possible - moving into a monopoly position and developing the communication and transportation network. The conditions that affect the sphere of circulation for effectively selling commodities in the market - the realization problem - include the development of coherent internal and external markets, an increase in the purchasing power of the population across regions, the continuous creation of new needs for individualized consumption, and the reduction of the turnover time of capital. Here again, capitalists act individually and collectively to improve these conditions, while the state and the local state play a crucial role in terms of securing minimum wages, welfare programmes, providing infrastructure, etc.

Finally, conditions that affect the sphere of reproduction involve, among others, the regulation of the capital-labour conflict, at the place of production and reproduction of the labour force. In the first case the capital-labour conflict is regulated through collective bargaining, control of unionization, labour legislation, welfare programmes for unemployment and/or direct oppression. In the second case, the reproduction of the labour force outside the factory is regulated through public and private intervention moving beyond its simple biological reproduction through wages and regulating its wider social reproduction as well (Poulantzas, 1974a; Tsoukalas, 1975). The latter includes the educational system and a number of services (public and private) which constitute the means of collective consumption at the urban and regional scale, such as transportation, schools, parks and recreation facilities, environmental protection, housing, health care, etc. (Castells, 1972; Lojkine, 1977). It is not what is necessary to ensure that workers will be capable of maintaining and reproducing their labour power; rather it is what is necessary to ensure that workers will be willing to accept their status as workers and to accept their mode of exploitation. An important role in the regulation of the capital-labour conflict at the place of production and reproduction is played by the wider ideological hegemony of the bourgeoisie (e.g. mass media, the educational system, the church). The creation of a 'good business climate' and the ideology of 'individualistic values' separated from politics are equally important conditions for high profits with increasing productivity or reducing labour cost, since they successfully mystify the conditions of capitalist exploitation and reduce class consciousness into consumerism and political apathy.

Although these fundamental concepts and processes are quite familiar, their contradictory implications within the geographical setting have seldom been explored. One reason for this may be that very often radical analysis has narrowly focused on one aspect of the problem. I argue, however, that such a partial viewpoint mystifies and obscures tendencies and contradictions that permit the production of surplus value to diverge from its geographical distribution (Harvey, 1982). Such an approach requires a fresh look at these localized conditions not as simple attributes of place but as 'active moments' in the geographical organization of capitalism.

The improvement of this whole tissue of antagonistic conditions in all sectors and regions of a social formation is the driving force behind capitalist development and expansion, and it is achieved through competition and state intervention. Through these improvements, general conditions of social production across regions and sectors tend to equalize towards a social average. As Marx observed in _Kapital_, Volume III, profits tend also to approximate the social average; in other words, a _tendency towards equalization of the rates of profit_ is in operation.[1]

The tendency towards equalization of the rate of profits tends also to produce an homogeneous social space, where the demands for individualized profitability, i.e. privatizing gains while socializing losses, destroys local, territorial cultural identity (Lefebvre, 1979; Friedman, 1977). Corporate and state expansion homogenizes urban and rural landscapes into subordinate aggregates, so that differences between the various '_topos_' (localities) are minimized, being evident only in different names and differences in travel time. And, as this tendency spreads over the national territory, it does so also at the international scale, though with certain limitations. But the results of capitalism remain the same: destruction of local _topos_ and their replacement with an homogenized network of specialized points of production, consumption and reproduction.

This tendency, however, creates problems for capitalist development. The very tendency towards profit equalization and spatial homogenization minimizes capitalist expansion, since no excess profits can be obtained with equal rates of profit in all sectors and all regions. No more accumulation will be possible except that necessary for demographic growth, and this itself would be modified in its own turn by the impact of the severe economic stagnation that would ensue (Mandel, 1975). Furthermore, the equalization tendency threatens the reproduction of capitalist social relations, which are founded on the unequal distribution of the social product both between capitalists and workers and among capitalists themselves.

A very effective counterbalance to the equalization tendency has been a combination of social practices, by individual capitalists, the state and local authorities towards _differentiation_ of the very conditions described earlier across sectors and regions. For example, the introduction by an individual firm of technical innovation could differentiate it from others and restore its profitability; a managerial strategy to increase wages in a certain firm or region above the national agreement with the state creates 'barriers of entry' to other less powerful firms in the same sector or pushes others to close production; an allocation of public investments to specific regions increases their competitiveness in attracting private capital vis-a-vis other regions; and the introduction by the state of a special land-use legislation or the construction of a new highway system assigns specific and differentiated roles to every territory, so that a parcelled and fragmented regional space is produced (Lipietz, 1977; Poulantzas, 1978). These observations help to clarify another tendency, namely that towards differentiation of the rate of profit.

The organization of space is a very important element in the tendency towards differentiation. Some spatial differentiation is inevitable in any mode of production, derived on the one hand from the simple friction of distance, and on the other from the basic principle of nodality and agglomeration in human spatial organization. These transhistorical spatial characteristics have been used by capitalists and the state to produce a disarticulated and fragmented regional and urban space, based on the relative immobility of labour and the relative mobility of capital.[2] The spatial disarticulation secures greater profitability in central versus peripheral locations and contributes to the regulation of capital-labour conflict through the fragmentation of working class neighbourhoods in cities and through the differentiation of urban versus rural districts in regional space (Lojkine, 1977: Lipietz, 1977). Thus, increasing differentiation in production and reproduction relations produces a disarticulated space and serves the function of restoring or maintaining conditions favourable for subsequent profitable accumulation. As Harvey (1982) observed:

> The homogeneity towards which the law of value tends contains its own negation in increasing regional inequalities. All kinds of opportunities then arise for competition and unequal exchange between regions ... The result is a chaos of confused and distorted motions towards both homogeneity and regional differentiation. (Harvey, 1982, p.441)

Taken together within the framework of geographically uneven development, these contradictory relations can be summarized and focused around the principal contradiction of spatial differentiation and spatial equalization in capitalist development. This contradiction is specific to the capitalist mode of production and central to capitalism's spatial problematic. Whether we refer to the internationalization of capital within a world system, to regional inequalities within a particular social formation, or to the political economy of urban space, geographically uneven development arises from these two simultaneous and opposing tendencies, and acts back upon them in terms of both the logic and strategy of capitalist accumulation and the logic and strategy of working class struggle. A principal conclusion from this discussion relevant for my argument on uneven regional development under the CMP can thus be summarized in the following contradictory relationships (see also Lefebvre, 1974: Palloix, 1972):

- On the one hand, there exists a tendency towards equalization of the conditions of production, the realization of surplus value, and of capitalist reproduction among sectors and regions in a given social formation, resulting in an equalization of rates of profit and the homogenization of regional space.
- On the other hand, there is also a tendency towards differentiation of the conditions of production, surplus value realization and capitalist reproduction among sectors and regions in a given social formation, resulting in the differentiation of profit rates and a disarticulation of regional space.

The broad significance of the contradiction has been elaborated by a number of Marxists. Palloix (1975) has been perhaps the most explicit analyst of equalization/differentiation on an international and regional scale, showing the connection of the contradiction with Lenin's 'law of uneven development'. Emmanuel (1969) and Mandel (1975) emphasize its economic implications, while Lefebvre (1972) puts his emphasis on the spatial dimension of the contradiction, especially at the urban scale. Lipietz (1975) discusses the 'perpetual struggle' between 'inherited' and 'projected' space in regional development. Finally, Poulantzas (1978) identifies the linked pairs of paradoxical oppositions such as fragmentation versus homogenization, division versus unification, and 'parcellization' versus structuring, in connection with the key role of the state. From a regional point of view the most important 'spatial barriers' for equalization of the rates of profits and at the same time the most important 'spatial aids' for excess-profit realization, can be organized in two large groupings: (a) those related to conditions of capitalized commodity production only, relevant for the discussion of uneven regional development in advanced capitalism, and (b) those related with the articulation between capitalist commodity production and simple commodity production, relevant for less advanced capitalist social formations.

In the first case, the most important of such conditions are the following: regional differentiation of the rates of exploitation and of the conditions of social reproduction of the labour force; regional differentiation in the organic composition of capital; regional differentiation in labour productivity; differentiation in turnover time of capital and in the realization process; and labour and capital mobility. Marx envisaged the possibility of this when he wrote:

> Capital succeeds in the equalization (of the rates of profit) to a greater or lesser degree, depending on the extent of capitalist development in the given nation, i.e. on the extent the conditions in the country in question are adapted for the capitalist mode of production. ... The incessant equilibration of constant divergences is accomplished so much more quickly (1) the more mobile the capital, i.e. the more easily it can be shifted from one sphere (of production) and from one locality to another; (2) the more quickly labour-power can be transferred from one sphere to another and from one production locality to another ... (K. Marx, Kapital, Vol.3)

In the second case, aside from the previous differentiated conditions, different conditions of accumulation in sectors and regions which do not all operate under capitalized conditions of production result in differences in the rate of profit among sectors and regions, because, as Marx argues:

> ... equilibration itself runs into greater obstacles, whenever numerous and large spheres of production not operated on a capitalist basis filter in between the capitalist enterprises and become linked with them. (K. Marx, Kapital, Vol.3, p.196)

These obstacles hinder the conditions of accumulation in various regions within a nation and acquire an even greater weight on the international level (Emmanuel, 1969). In the periphery and semi-periphery the greater relative immobility of capital, the prevalent immobility of labour and above all the complex combination of various forms of production within a social formation under the dominance of the CMP, are the fundamental factors that account for differences in the rate of profit. The point is that, during its effort to dominate all other modes of production and through relative improvements in both capital and labour mobility, the CMP is moving towards an equalization of the rates of profit. But the obstacles discussed above and the nature of capitalist competition counterbalance this tendency. As Mandel (1975) argues:

> ... the differences in the level of profit arise out of the competition of capitals and the inexorable condemnation of all forms, branches and areas which fall behind in this race and are thus forced to surrender a part of their 'own' surplus-value to those in the lead. What is this process, other than the continual production of underdeveloped firms, branches, areas and regions? (Mandel, 1975, p.85)

The argument here is not simply that capitalist development is organized unevenly over the regions of a social formation, for some regional unevenness is the result of every social process. Modern capitalism, however, actively creates, intensifies and seeks to maintain regional differentiation, while simultaneously generalizing a tendency to increase equalization and to reduce regional differences. In this way, the production of homogenized and disarticulated regional space (through the actions of individual capitalists and the state) contributes to the reproduction of capitalist relations of production. This dialectical tension between equalization and differentiation is the underlying dynamic of uneven regional development.

What Mandell and others fail to conceptualize, however, is the fact that the movement towards equalization or differentiation is not a matter of free choice by capital and/or the state. It is the outcome of time-space specific <u>struggles</u> between capital and labour over the production of surplus value, and struggles between capitalists over the distribution of surplus value. The reproduction of the general social conditions is an 'active moment' within the overall accumulation process, and is bound up with the perpetual motion in the spatial configuration of capitalist social relations. Thus the reproduction of these conditions becomes a political necessity via the intervention of the central state and the local government (Lojkine, 1977; Preteceille, 1981; Folin, 1981). This is why issues such as capital restructuring, regional and urban declines and 'the regional crisis' are part of the crisis of the state in virtually every European capitalist social formation.

4.2 DIFFERENTIALLY LOCALIZED ACCUMULATION

The analysis of the principal contradiction of equalization/differentiation has shown the dynamic of uneven regional development. It has not shown, however, the underlying tendencies through which it gets into motion. I argue that the underlying tendency is based on what I shall call differentially localized accumulation (DLA) and thus represents the spatially differentiated impulses towards capitalist expansion and realization of super-profits. The end result of the operation of the tendencies is what I propose to call the geographical transfer of value (GTV). GTV represents the reallocation of part of the surplus value produced by workers in the peripheral sectors and regions of a social formation to the pockets of capitalists in the core, whose firms thus acquire a number of characteristics such as high capital intensiveness, monopoly, higher wages, shorter turnover time of capital, etc.

The differentially localized accumulation in a region of a social formation depends first on the historical mode of integration of the region concerned into national and international division of labour (IDL); secondly, on the structure and current performance of the producing sectors in the region, in terms of jobs, productivity and incomes; thirdly, on the role of the state and local authorities; and fourthly, on the practices of civil society.

The specific historical mode of integration of a region into the IDL depends on the means of capitalist penetration of the region, a penetration which transforms incompletely capitalized communities from relatively self-sufficient organizations for the production of use values into specialized and dependent units, producing for the national and international market. Thus, current attempts at regional planning must seriously take into consideration the region's historically established DLA, because the social and spatial structures that have been developed in the course of capitalist penetration are reflected today as either obstacles or advantages for further capitalist development of specific social groups and localities.

For example, historical research in southern Europe has shown that the mercantile era (16th to 18th centuries) was shaped by a specific organization of the Mediterranean market, i.e. by early agricultural specialization, long-distance trade, the mode of labour control and the power of the urban bourgeoisie (Braudel, 1972). These factors influenced accumulation processes which, in turn, were spatially differentiated, resulting in a core-semiperipheral-peripheral socio-spatial structure (social and spatial division of labour). Therefore, new investments in economic activity during the 19th and early 20th century were spatially distributed according to the needs of production and took advantage of already established uneven socio-spatial structures. The new investments reproduced the core-periphery structure, though changing its particular qualities and intensities.

This interaction is very important. First, it shows that not only do production and accumulation shape regional space, but also that the pattern of established spatial differentiation influences the course taken by accumulation in later historical periods. Secondly, it shows that the degree of attractiveness of a particular region to the dominant economic activity is a function of both its historical past and its present features of production, including expected rate of profit.

The location of an economic activity in a particular region, however, does not necessarily contribute to accumulation processes in that region or to its own regional development (Palloix, 1975). This is precisely the essence of the theory of accumulation on the world scale (Amin, 1974): accumulation is taking place independently of the particular local distribution of economic activities. Most Marxists, however, had ignored the geographical dimension of accumulation and concluded that any attempt to introduce space into this process was misleading.[3] But accumulation on a world scale does not happen _en l'air_; accumulation is an abstract but at the same time concrete process. It has to happen somewhere; it needs not only labour, means of production, raw materials and a set of social relations, but also a _topos_, a territory. To understand the relationship between accumulation in the abstract and the various _topos_ of a social formation (the nation, regions, settlements, factories and rural districts, etc.) we must analyse the process of accumulation itself, i.e. the creation, circulation and realization of surplus value placed in a concrete spatial context.

The process of capital accumulation in a region consists of two movements which, as I argued before, are separated in time and space: first, the appropriation of surplus value in the sphere of production, which means that workers produce commodities and are compensated at a rate that represents less than their contribution to output; and secondly, the realization of surplus value in the sphere of circulation. Profits for individual capitalists are calculated at the end of the second movement.

In this respect, the DLA in various regions of a social formation will increase the greater the mass of surplus value produced and the greater the proportion of locally realized surplus value. Economic sectors in these regions enjoy high rates of accumulation and a high rate of economic growth. Similarly, a low DLA for a region will mean either that low quantities of surplus value are produced by the sectors located in this region and thus not enough surplus value exists for extensive realization; or that part of the surplus value produced cannot be realized locally. In either case, the rate of accumulation will be low, followed by low rates of growth. In the case of the partial realization of surplus value in the region where it has been produced, this non-realized surplus value is not lost in thin air. It has to go somewhere, and this 'somewhere' involves a sectoral and geographical transfer, usually from peripheral to core sectors and regions. There exist, however, cases where losses of value could also take place. I will extend this argument in the next section. Meanwhile, I note that possible mistakes in production choices, bad management or a sudden drop of value of the means of production could end in losses of value and a waste in the allocation of the social labour (De Vroey, 1981).

Starting with the first movement of accumulation, the regional production of surplus value (both absolute and relative) will depend upon the conditions influencing the sphere of production analysed in the discussion of the contradiction of equalization/differentiation. Take, for example, one of these conditions, the organic composition of capital. A region will generally have a greater DLA when its productive structure operates with an average organic composition higher than that for other regions. This is largely explained by the fact that higher organic composition increases total productivity and output and relatively reduces the vulnerability of capitalists to a demanding and unionized working class. In addition, the distribution of profits among capitalists through competition and the tendency towards an average rate of profit is organized according to one's holdings of constant capital and not to the corresponding amount of labour time involved in production (Bauer, 1924). Furthermore, if the DLA in a region were dependent entirely on maximizing absolute surplus value, then firms located there would have no incentive to increase their organic composition and to improve methods of production through technology and product innovation. This is precisely what all capitalists tend to do: to improve their own conditions of production and to move from absolute to relative surplus value production.

As the organic composition of capital rises, however, there is a tendency for the rate of profit to fall unless the rate of exploitation increases sufficiently. In the long run, increases in the rate of exploitation cannot completely counteract the rising organic composition, and only an effective differentiation of the other social and material conditions discussed previously could temporarily prevent crisis. And, when the decline in the rate of profit in a particular region becomes sufficiently serious and can no longer be compensated for by existing social relations and institutions, the least profitable firms go bankrupt and capitalists increasingly withhold investments because there are no profitable outlets.

The second movement of capitalist accumulation, the realization of surplus value in the sphere of circulation, is more complicated. While production is primarily organized by every capitalist individually, realization is more dependent on expanding socialization of consumption, thus demanding first the intervention of the state and secondly those practices of civil society that help socialization of consumption. The state attempts to maintain capitalist production relations by moderating, reforming and transforming the relations of exchange. Civil society embraces widely divergent practices, from family relations to commodity markets, from trade union organizations to religious bodies and regional movements. It is the combination of state's and civil society's practices that makes a particular region 'attractive', not simply for initial investment but also for spending money and re-investment of profits. From a realization point of view, the DLA for a region will be greater if all surplus value produced is also locally realized. This does not mean that all commodities have to be sold within the region, but that the net profit gained from the production and selling of commodities must, in the final analysis, be accumulated in the region where the production is organized. Consequently, the DLA will be lower for a region when part (or all) of the surplus value produced is not locally realized.

In the first case, the conditions for a high DLA are increasingly dependent on the existence of what mainstream regional theory calls a 'developed regional market' within the region and/or an expanding demand for region's exports. In this respect, regional theory reduces the practices of both the state and civil society to simple economic indicators of aggregate demand. But, while mainstream regional development theory fails to see the distinctive character of capitalist production and the state, conventional Marxism tends to consider social formations as though they constituted relations only of production and the state (Holloway and Picciotto, 1978). Little attention is given to the processes by which agents get constituted as subjects outside the social relations of production and the state. And no space is allocated to social classes, apart from their determination by the economy or their effects within the state. Finally, and as a result of the previous two points, such Marxism cannot analyse in a non-reductionist way social movements around specific issues such as gender, ecology, region or nationality, which may influence the realization process.

Recalling the previous discussion on the general conditions that affect the sphere of circulation, I note that the more these conditions are advanced, the greater the DLA in terms of realization. These conditions are particular to a place and are both an historical product of past accumulation phases and a product of the current performance of region's economic structure and the practices of the state and local institutions. The discussions of the contradiction of equalization/ differentiation has shown, however, that capitalist development itself creates barriers and hinders the development of these factors. For example, there exists an intrinsic contradiction between the production of surplus value and the conditions of the realization of surplus value. For realization not to be a problem, the growth of aggregate demand must occur at the same rate as the growth in surplus value (Sweezy, 1956). This is always problematic in capitalism, because individual capitalists tend to minimize their wage bill, thus restricting the development of effective demand on the part of workers. In the absence of new sources of demand (e.g. from the state, from abroad, etc.), capitalists will experience a fall in the actual rate of profit.[4]

Another example is related to the reduction of the turnover time of capital, a very important element in the realization process, what Harvey (1975), following Marx, calls the 'annihilation of space by time'. In the effort to reduce the turnover time, capital tends to be geographically concentrated, and this reduction in realization and circulation costs, helps to create fresh room for new capital accumulation. But this tendency, which is the driving force for capitalist urbanization and agglomeration in certain growth points, creates an immobile physical landscape (e.g. cities, transport systems, harbours, factories, etc.) which, as Harvey emphasizes:

> ... is both a crowning glory of past capital development and a prison which inhibits the further progress of accumulation because the very building of this landscape is antithetical to the 'tearing down of spatial barriers' and ultimately even to the 'annihilation of space by time'. (D. Harvey, 1975, p.273)

The resolution of the contradiction is always a knife-edge path between the destruction and preservation of the historical landscape, which has become inefficient for capital under the pressure for the creation of a new, and more capital efficient, landscape which, after one or two generations, may itself face similar problems.

Thus, taking into account the accumulation process as a whole, three major factors determine the DLA in a region. First is the historical mode of integration of the region concerned into the national and internationl division of labour. But, while past accumulation could be an historical advantage for current performance (and vice versa) this is not always the case. It depends - and this is the second factor - on the current performance of the regional economic structure and on those general conditions that affect the sphere of production. The more capital intensive and the more technologically advanced is the regional economic structure, the higher its DLA potential. And thirdly are the practices of the state and civil society that influence the sphere of circulation and the reproduction of labour force. The more the state and local authorities intervene to increase socialized consumption and the more the practices of private and/or collective agents coincide with this aim, the higher the DLA potential in this region. Under these conditions accumulation could take place within a region, with all reservations of what 'within' means. In this case, capitalist expansion is expected, while in certain regions that fail to secure these conditions, DLA remains low and capitalist expansion declines. In both cases, however, uneven regional development is reproduced.

4.3 THE GEOGRAPHICAL TRANSFER OF VALUE (GTV)

Considering the various regional DLA and corresponding differences in regional development, the argument of GTV runs as follows: regions are open systems having relationships with national and international markets; the contradictory interplay between equalization and differentiation portrays the dynamic and the competitiveness of these relationships resulting in DLA; thus, if a sector or a firm located in a certain region cannot realize part of its produced surplus value (apart from losses of value), then this non-realized value is transferred and realized elsewhere, by another sector in another region. This process, in which part of the surplus value is drained away, could result on the one hand in a slow-down of accumulation rates in the value-losing region (negative GTV), and on the other in an acceleration of the accumulation rate in the value-gaining region (positive GTV). The latter, however, is contingent on the practices of individual and collective agents in terms of the use of the transferred surplus. As Gore (1984) argues, criticizing Baran (1957), there is no reason for concluding that, if surplus were larger or smaller in a region, then growth or stagnation follows. Without consideration of the way the surplus is used within the region, the conclusion is merely a hypothetical assumption.

The geographical transfer of value could be seen as adding a spatial component to the sectoral transfer of value analysed by Marx in the context of his general law of value and the notion of socially necessary labour time. What is more important, however, is the introduction through GTV of the dynamic changes in regional structures resulting from the practices of the state and civil society. In this context it is not the only, or even the major, factor influencing or explaining uneven regional development, but one among many.

In mainstream regional theory Myrdal (1957) with his notion of 'backwash' effects has come close to the concept of GTV, inasmuch as the growing regions attract factors of production away from other regions, leaving them '... more or less in a backwater'. Myrdal's idea means that the benefits to core regions will be at the expense of development in factor donor regions. The process is cumulative in the sense that an initial historical push or pull of factors to one region and not to another will tend to move factors increasingly towards the core regions and away from peripheral ones. Similar arguents have been developed for Italy by Sylos-Labini (1965) and Downie (1958) that there is a tendency for more efficient firms in core regions to grow at the expense of less efficient ones. Downie calls such a process a 'transfer mechanism' and states that:

> ... if the transfer mechanism continued to be operated by an unchanged set of relative efficiencies the ultimate result could only be the concentration of the whole output of an industry in the hands of one, the most efficient firm. (J. Downie, 1958, p.60)

A similar view is adopted by dependency theorists. The fact that the centre expropriates the economic surplus of its periphery for its own use is central to the transfer of surplus development thesis with all its deficiencies analysed in Chapter 3. The importance, however, of dependency theorists' contributions is that they allow one to think in terms of an area's surplus (see Baran, 1957). But their analysis does not avoid two important fallacies. First, one must not conclude that the surplus of a region is extracted by another region, as if regions are exploiting other regions, which is meaningless. Secondly, one cannot think of a region's surplus independently from sectors and firms existing in the region. The concept of GTV instead proposes to investigate the transfer mechanism among or within sectors which are located in different regions. The latter is promoted through the analysis of the accumulation process and the law of value, a task which both the backwash-effect and dependency theorists failed to accomplish.

A version of this alternative view, based on the labour theory of value, has been adopted by unequal exchange theorists (see Emmanuel, 1969; Amin, 1977; Mandel, 1975).

The geographical transfer of value is often equated with unequal exchange (Walker, 1978; Webber, 1982; Browett, 1984). This is not correct, however, because GTV, without denying its connection to unequal exchange, is based on a much broader concept including various forms and channels of value transfer. In this respect, unequal exchange is one form of indirect GTV which includes also unequal trade, unequal rewards to labour and other forms (see De Janvry, 1981). I will have the opportunity to discuss these forms of GTV in detail in Part II, taking agriculture in southern Europe as an example. Here I want to present briefly the essence of the theory of unequal exchange first introduced by the Greek Marxist economist, Arighiri Emmanuel, in 1969: selling commodities below value and buying others above value. Unequal exchange is a specification of the general transformation problem and the transfer of value, both sectoral and geographical, in terms of trade and direct exchange between two specific sectors and regions.[5] It has been applied first at the international scale, but Emmanuel himself and later Lipietz (1977), Sayer (1977) and Liossatos (1979) examine the usefulness of this approach also at the regional scale.

The publication of Emmanuel's book was followed by an extended debate among many of the protagonists of current Marxist development thinking, including C. Bettelheim, S. Amin, C. Palloix, O. Braun, P.B. Braun, M. Kidron, E. Wallerstein, J. Kay and

E. Mandel.[6] Emmanuel found himself arguing against both Marxists and bourgeois theorists. Marxists had claimed that appropriation of profit from investment in underdeveloped countries was the central mechanism of exploitation. Bourgeois theorists claimed that underdevelopment was due to insufficient production for capitalist markets (P.B. Brown, 1974). Emmanuel instead proposed unequal exchange as the major mechanism of exploitation directly applying the law of value in international relations.[7]

In short his argument ran as follows: in a capitalist economy all factors of production are not fully and equally mobile - not only capital and labour but any direct or indirect component of production. Thus, each factor cannot receive a standard return whenever or wherever it is put to work. This is true at the international scale but also, perhaps to a lesser degree, within a nation. Thus the distribution of surplus-value among classes, social groups and class fractions which control these factors is not the same everywhere. It varies from region to region in accordance with the relative strength of the classes or groups in control of the less mobile factor. In this respect, the mobility of factors of production follows the principal contradiction of equalization/ differentiation. An uneven distribution of the socially produced surplus value within the system is thus set up, with a transfer of surplus value from one class to another and from one region to another via the equalization of returns to the more mobile factor or factors. Since, in the CMP, capital is more mobile than labour, the socially produced surplus value in a social formation will be shared among the capitalists in proportion to the amount of capital that each has invested and not in proportion to the labour spent to produce the commodities. Under such conditions, when two sectors exchange their products and the first has a higher organic composition of capital (which usually means higher productivity) than the second, the exchange is an unequal one; it is an exchange of less against more labour, which inevitably leads to a transfer of value from the second sector to the first or, in the case of their geographical separation, from the second region to the first region. This is because, as Emmanuel argues:

> ... behind the comparison of commodities lies hidden a comparison between the different

labours needed to produce them (p.210)

As I pointed out in the previous chapter the main problem with these accounts of uneven development was their over-emphasis in the sphere of circulation and exchange, the unclear definition of exploitation via 'external' factors; the assumption of an equilibrium in capital market; and the introduction of unequal exchange as the only or the major mechanism of value transfer. These overstated arguments and the lack of any empirical investigation made unequal exchange highly vulnerable for a reliable explanation of uneven development. As Sayer (1977) argues, too much debate on the theory has been conducted without adequate empirical research to supplement abstract comments (Sayer, 1977, p.40).[8]

In the late 1970s and early 1980s empirical and theoretical research around the value transfer concept shifted the emphasis to concrete investigations. Liossatos (1979) for example gives a mathematical explanation and demonstrates that unequal exchange on a regional scale exists and transfers value. He uses Morishima's formalization of Marx's economics combined with Straffa's theory of prices of production. His method is based on sectoral disparities which are in turn translated into regional inequalities by giving the economy a multi-regional structure. Thus, it is the spatial division of labour that translates intersectoral to interregional unevenness and not any 'intrinsic attributes' of space. He concludes that:

> ... some sectors and regions of high capital intensity <u>and</u> wages (relative to the national averages) realize more income than the monetary expression of the value they create. (Liossatos, 1979, p.28)

Following Morishima's and Liossatos' formulations, Marelli (1983) provided a most useful analysis of intersectoral and interregional transfer of value in Italy, using input-output data. His method makes it possible to calculate the values of commodities and subsequently the rate of surplus value, the organic composition of capital and the value rate of profit for each sector and for each region. Denying the fallacy of exploitation of one region by another, he assumes that the less developed regions of a given country are specialized in sectors with low value composition of capital and/or high rate of surplus value and/or low organic composition of labour. His evidence is clear that transfers of value do take place but this, he concludes can provide only 'a partial explanation for the phenomenon of regional disparities' (Marelli, 1983, p.68).

Finally, Webber and Foot (1984) provide an elaborated empirical measurement of unequal exchange at the international level using input-output data for Canada and the Philippines. Their method is similar to those of Marelli and Morishima, measuring commodity values by the average amount of socially necessary labour it takes to produce them. Their empirical results also strongly support the hypothesis that unequal exchange transfers value from less developed countries to more developed ones by foreign trade.

The findings and empirical results of these accounts of uneven regional development are very useful for my analysis of GTV. They remain, however, within the limits of unequal exchange, which, as I noted already, is only one among the many forms and channels of GTV. This limitation did not allow for a further analysis of the realization process and its spatial variations. Finally, no attempt is made to connect the transfer of value with the role of the state, local state and the various forms of struggle. My focus on these issues, including the analysis of the contradiction of equalization/differentiation and the DLA, serves to emphasize the particular importance of space-time structures and actions for the realization of surplus value. Next, I consider how sectoral and regional variations of such issues may give rise to GTV (Hadjimichalis, 1980, 1984).

Marx defined the realization of surplus value in terms of the successful movement of capital through the three phases of circulation: value appears first as money (when capitalists are acting as buyers); secondly, it appears as labour force (when capitalists are acting as organizers of production); and thirdly it appears as a material commodity (when capitalists are acting as sellers). These transformations are only by exception taking place at the same time and very rarely within the same territory. This means that realization of surplus value occurs within or outside the area where it has been produced, and in short or extended periods of time. Now, two crucial points must be clarified. First, the timing and the place of the realization process is dependent on the conditions influencing the sphere of circulation. In other words, GTV as a hidden tendency within the geography of capitalism is the 'external' outcome of the practices of the state and civil society. Secondly, the 'internal' pre-conditions for GTV to occur are largely based on the sphere of production, in other words on the practices of individual capitalists. I have to emphasize that neither value nor surplus value are created in circulation, because in this process commodities are merely exchanged, not created. It is in circulation, however, that the value magnitudes take their money forms, value the form of money-price, and surplus value the form of money-profit (Morishima, 1979). The sum of profit for individual capitalists located in a particular locality are calculated at the end of the third movement. They form, as I have already pointed out, a 'localized accumulation'.

From a regional point of view the GTV operates between two or more regions that have different DLA through direct and/or indirect means. Direct GTV occurs when certain actions by social agents (usually but not exclusively outside the donor region) mediate directly so that part of the surplus value produced locally is transferred elsewhere. Indirect GTV occurs when the same process operates without direct mediators, simply through what orthodox economics call the play of the market forces. Behind this, however, lies, as I remarked already, the sum of concrete practices by individual capitalists, the state and civil society. Both direct and indirect means of GTV are dependent on the historical mode of integration of the region concerned into the national and international division of labour.

Direct GTV

The GTV, through direct means, involves a series of actions which vary historically from direct violence, war and plunder, to more sophisticated policies, such as state taxation policies, public transfer-payments, multinational profit repatriation and transfer-pricing, the net export of fees and royalties over 'public aid' receipts, the transfer of direct control over peripheral capital and natural resources to multi-regional and multinational capital, investment of the local ruling class profits outside the region, followed by working class savings transfer to urban areas via the banking network, the 'brain drain', etc. Other direct means of GTV, price fixing and manipulation of terms of trade, can be better understood within the wider framework of indirect GTV and price formation discussed below.

These means of GTV are mediated either officially by the state and the local state or by international industrial and financial capital, and by local ruling classes which, in the case of southern Europe, tend to invest in real estate in core urban regions. Under such conditions only part of the surplus value produced in the region remains to be locally realized, so that the size of the regional market and accumulation is distorted and decreased accordingly. In contrast, accumulation in core regions - within or outside the country - increases through the realization of geographically transferred surplus value. This direct GTV has received detailed attention in the contemporary literature, both for its international and regional implications.[9] In the Mediterranean it has been studied in terms of state action (Pastore, 1966; Palloix, 1975; Hadjimichalis, 1976; Holland, 1976; Secchi <u>et al</u>., 1978).

For example, in southern Europe export-oriented or assembling industries that sell to multinational firms seem to be one of the designed futures for peripheral region industrialization (via growth pole strategies). These 'export processing zones' among other things leave the host region/ country with a minimum bargaining advantage. Not only is the export/processing manufacturing activity extraordinarily 'footloose', dependent as it is on neither local resources nor local markets, but it is also likely to bind the host region both to sources of inputs and to market outlets over which it has little if any control. The result is a selective spatial integration on the world scale where a specific plant and location is integrated into the international market.[10]

Direct GTV, without being named as such, has been the analytical focus of many dependency and neo-populist theorists. It has been used as a major explanation of underdevelopment due to 'external' factors, i.e. the penetration of advanced capitalist economies to peripheral ones. In these cases attention to 'internal' factors, such as the structure of local production, social classes, the role of the state, the practices of civil society, etc., attracted limited or inappropriate attention. For these reasons it has attracted fair criticism from those who argue that what is important is the articulation between external and internal factors and not the a priori dominance of one over the other.

Indirect GTV

One could argue that direct GTV can be controlled and possibly eliminated through state intervention and deliberate regional policies. But even in the ideal case of its complete control there will always remain the GTV through indirect means, based on the operation of the capitalist market. In other words, the process of accumulation facilitates the transfer of value from one sector and region to another via a number of channels such as unequal trade, capital and labour mobility, unequal exchange, unequal rewards to labour, etc.

The indirect transfer of value is directly associated with the transformation of values into prices of production and then to actual prices in the market, known as the 'transformation problem', a debated and complex issue in Marxist political economy and one which has also attracted criticism from non-Marxist economists as well.[11] To a lesser degree, it is also associated with the theory of unequal exchange (Emmanuel, 1969; Amin, 1977), which should be seen as part of the general transformation mechanism described by Marx in Kapital, Volume III, and not a substitute for it, as some critics argue.[12]

Consider, for example, a six region economy, with each region's dominant sector specializing in the production of certain commodities or in the provision of certain services.[13] Each sector in each region will have different conditions of production in terms of organic composition and rates of exploitation. The situation can be depicted as shown in Table 4.1.

In this case, other conditions being equal (e.g. turnover time, monopoly, cost of reproduction of labour force, etc.), the rate of profit moves in the opposite direction to the organic composition of capital. For obvious reasons, Marx claimed, there must be a tendency towards a general rate of profit in the economy. If this were not the case, all capital would inevitably flow to that sector and region where the rate of profit was highest - in my example Sector V, in Region R5 - which would mean on the one hand excessive production of commodities in this sector, and on the other a shortage of commodities and services produced and offered in other sectors and regions. In this respect, the tendency towards equalization of the rates of profit arises from the need of social reproduction of

Table 4.1

Sectors and Regions	Constant Capital C	Variable Capital V	Surplus Value S	Rate of Exploi- tation $\frac{S}{V}$	Organic Compo- sition $\frac{C}{V}$	Rate of Profit $\frac{S}{C + V}$
I_{R1}	80	20	10	50%	4.00	10%
II_{R2}	65	35	30	85%	1.85	30%
III_{R3}	70	30	30	100%	2.33	30%
IV_{R4}	95	5	5	100%	10.00	5%
V_{R5}	55	45	40	88%	1.22	40%
IV_{R6}	60	40	35	87%	1.50	35%
Total Social Capital	425	175	150	85.7%	2.42	25%

capitalist relations of production. This is achieved through struggle, and the discussion of the contradiction of equalization/differentiation has shown how this leads to crises. The existence of this general rate of profit, however, is inconsistent with equivalent exchange, considering the different conditions of the production of surplus value in various sectors and regions. If all sectors in all regions did sell their commodities at equivalent prices in relation to embodied labour time and were thus able to realize all their produced surplus value, then the sectors with the lower organic composition, lower wages, longer turnover time of capital, less monopoly control, etc., would show a higher final profit (in production and realization), a result which is inconsistent with most essential features of capitalist development.

As Kay (1975) argues, many economists have considered this inconsistency so fundamental as to invalidate the law of value entirely. However, as he and others (Meek, 1956; Emmanuel, 1969; Amin, 1977; Koshimura, 1975; Shaikh, 1980) have shown, the problem can be resolved with the observation that commodities do not exchange at prices equivalent to their values and embodied labour time. In Marx's words:

> ... the price of production of a commodity may lie above or below its value, and coincides with its value only by way of exception. (Marx, Kapital, Vol.3, p.758)

The law of value states that the prices of all commodities produced in a national market taken together must be equivalent to their collective value, but the price of an individual commodity is not necessarily equal to its value. The tendency towards equalization of the rate of profit does not equalize the price of commodities. On the contrary, the average rate of profit and the market prices permit a certain sector which operates in more favourable conditions to realize - aside from its own surplus-value - part of the non-realized surplus value of other sectors. As Marx states:

> It is therefore wrong to say that competition among capitalists brings about a general rate of profit by equalizing the prices of commodities to their values. On the contrary, it does so by converting the values of the commodities into average prices, in which a part of the surplus value is transferred from one commodity to another. (Marx, Theories of Surplus Value, Vol.II, Part I, p.30)

This sectoral transfer of part of non-realized surplus value over space comprises the indirect GTV, since production, realization, consumption and distribution are not taking place on the head of a pin.

To understand this indirect GTV in more detail, I will introduce, following Marx, two additional concepts. The first is the cost price

$$(P) = C' + V$$

where C' is the part of constant capital used in the production process. Cost price is not the value of commodities

$$= C' + V + S$$

produced in a day in the sector, but the value that has to be advanced in production (see Emmanuel, 1969; Kay, 1975). The second concept is the price of production: the price at which commodities are actually sold in the market and which is equal to the cost price plus the general rate of profit in the economy as a whole, achieved through the tendency towards equalization of rates of profit. C,

V and C' are percentages, the first two expressing the ratio of constant to variable in total capital, and the third, C', expressing the percentage of used to total constant capital C in the production process. Introducing cost price, the value of commodities and the price of production into Table 4.1 we have Table 4.2.

Table 4.2 also shows that deviations of values from prices balance out one another in the economy as a whole, since the sum of prices of production equals the sum of values of commodities. Some sectors and regions, however, did not realize all their surplus value - in my example II_{R2}, III_{R3}, V_{R5} and VI_{R6}. The total loss of value-losing regions was equal to 35 units of value. This 35 units of non-realized surplus value has been transferred to the value-gaining sectors and regions - I_{R1} and IV_{R4} - through the selling of their commodities at above their actual value. In this respect an indirect GTV is taking place from the four value-losing sectors and regions (II_{R2}, III_{R3}, V_{R5} and VI_{R6}) to the two value-gaining sectors and regions (I_{R1} and IV_{R4}).

In indirect GTV, the transformation from cost prices to prices of production does not involve any real change for the system as a whole (Morishima, 1979; Shaikh, 1980). The total mass of commodities, and the relative portions of it going to each class remain the same as before. By the same token, so do the sum of values and the sum of surplus-value. All that changes is the manner in which given production relations are manifested in circulation (Shaikh, 1980). What the transformation brings about is a different division of the total pool of surplus-value among individual capitalists (having their firms in different or in the same regions). With the cost prices, each capitalist realizes an amount of money-profit equivalent to the surplus-value contained in the commodities being sold. With prices of production, each regional sector's money-profit is no longer proportional to its surplus-value produced locally; since the sum of values (and hence the total surplus value produced in the social formation as a whole) is still the same as before, the above change of form has the effect of <u>redistributing</u> surplus-value from one sphere of production to another and from one locality to another. Thus, final profits are distributed among the capitalists in various sectors and regions according to each one's holding of capital (thus reflecting higher organic composition), for, as Marx

Table 4.2

	C	V	C'	S	Value of Commodities C' + V + S	Cost Price C' + V	General Rate of Profit R	Price of Production C' + V + P	Deviation from Value to Prices
I_{R1}	80	20	50	10	80	70	25%	95	+15
II_{R2}	65	35	55	30	120	90	25%	115	- 5
III_{R3}	70	30	60	30	120	90	25%	115	- 5
IV_{R4}	95	5	30	5	40	35	25%	60	+20
V_{R5}	55	45	30	40	115	75	25%	100	-15
VI_{R6}	60	40	35	35	110	75	25%	100	-10
Total Social Capital	425	175	260	150	585	435	25%	585	+35-35=0

pointed out:

> ... although in selling their commodities the capitalists of the various spheres of production recover the value of the capital consumed in their production, they do not secure the surplus value, and consequently the profit, created in their own sphere by the production of these commodities ... the various capitalists are just so many stockholders in a stock company in which the shares of profit are uniformly divided per 100, so that profits differ in the case of the individual capitalists only in accordance with the amount of capital invested by each ... (Marx, Kapital, Vol.3, p.158)

These value transfers could be better understood in terms of circulation profits, circulation losses and losses of value (De Vroey, 1981). Profits and losses in circulation result from the lack of full realization of the norms of exchange. They occur in different localities when there is a situation of seller's power to impose a premium on the equilibrium price, derived from a local strength, in terms of class structure, labour market and general conditions of production. The specific matter of these transfers is that globally they are cancelled out, since the circulation profits of some are, by definition, the losses of others; their sum amounts to zero.

Losses of value arise from two sources: lack of sale, or devalorization of the means of production (Harvey, 1982). The first results from mistaken choices of production and bad management in circulation. The second results from a sudden drop of value of the means of production. In contrast to losses in circulation, losses of value always represent waste in the allocation of the social labour of a commodity system. In this case, GTV does not occur; it is the 'price' paid for the way social cohesion is formed in a particular society.[14]

So far I have discussed certain aspects concerning the sphere of production which may give rise to GTV, such as differences in the organic composition of capital, in prices of production, in the rate of exploitation, and so forth. Nevertheless, there are other aspects as well which may give rise to GTV. These aspects have been called by Emmanuel (1969) 'other factors'; they are territorially defined, acting as a source of a kind of 'quasi-rent' for the region's DLA, such as special tax exemptions, price fixing, tariffs and terms of trade. These factors are not only direct means of GTV but are also applicable through the transformation of values into prices of production and then into actual prices in the market. What Emmanuel and other unequal exchange theorists do not include in these other factors - which I believe is crucial in understanding the role of regional space in uneven regional development - are the issues discussed previously in terms of the 'general conditions' influencing the spheres of circulation and reproduction. And especially those conditions that reproduce the geographical differentiation in the three phases of circulation (money, labour force, commodities) and the differentiation in the cost of reproduction of the labour force.

This is particularly important in cases where a regional equalization of organic composition of capital is in operation (e.g. in an advanced capitalist economy), and a tendency towards equalization of regional wages has been achieved through class struggle and state legislation. I argue that in these cases differential distribution of the general conditions of circulation and reproduction, such as the means of collective consumption and means that help to reduce the turnover time of capital (e.g. communication and transport systems, spatial concentrations of the tertiary sector, special services to capital, etc.), operate indirectly to influence the transfer value from peripheral to core regions. These other factors, as Emmanuel has shown, are <u>added as a coefficient to the cost price</u>, thus increasing the deviation of prices of production from values. A similar argument is presented by Lojkine (1977), who observed that regional and urban services of insufficient quantity and quality increase the cost of reproduction of the local labour force, thus entering indirectly into cost price formation. He emphasizes the fact that this is a net 'gift' to monopolies from the state, the former being able to take greater advantage from positive externalities of urban and regional development, due to their large multi-locational operations in production and distribution and their large number of employees. Finally, the idea of indirect GTV and the role of 'other factors' (including space), resembles the proposition of Harvey (1975) about the role of urban space in the redistribution of real income. Harvey discusses certain 'hidden mechanisms' -

which, if analysed further, could lead to the surplus value concept - that transfer income from one neighbourhood to another through the price formation of 'urban rent' (indirect means) and through the 'political and planning process' (direct means).

Table 4.3 uses the previous example of six region - six sector economy, but assumes now an equal rate of exploitation (100%) in all regions and sectors. It also assumes 'other factors' (F) to exist in the economy acting as a source of 'quasi-rent' to be added as a coefficient to the cost price. We could take as an example two groups: means of collective consumption (e.g. hospital beds) and means to reduce the turnover time of capital (e.g. banks). These other factors exist in the economy as a whole and also exist in every region unevenly distributed as percentages of the total volume of means of collective consumption or means to reduce the turnover time of capital. For example, the means available locally to Sector I in Region R1 to reduce its turnover time of capital equals 25% of the total means available in the entire country, while Sector V in Region R5 has available only 5%, etc. Similar examples could be used for the means of collective consumption at the regional scale.

Comparing Tables 4.2 and 4.3 we can see that in Table 4.3 differences in the deviation (e.g. ratio of extreme deviations) from values to prices due to other factors are greater than in Table 4.2, despite the fact that direct wages in the second case are equalized across regions and that the ratio of commodities to cost price in the second case slightly decreases. These differences also support the hypothesis of the DLA and the GTV in terms of a developed versus non-developed regional market, since a regional market's development is directly related to the means of reduction of turnover time of capital and means of collective consumption.

Finally, a major issue in indirect GTV (and to some extent in direct GTV as well) is concerned with cases where the labour theory of value is not valid, i.e. with cases where pre-capitalist forms of production are articulated with the capitalist mode of production (Rey, 1973). In this case the transferred surplus cannot take the form of surplus value but that of <u>surplus labour</u>. The magnitude and direction of transferred surplus labour are determined by the balance of forces among the CMP and the various pre-capitalist forms. In those cases the role of the state and civil society

Table 4.3

	C	V	C'	S	Value of Commodities C' + V + S	Other Factors F	Cost Price C' + V + F	General Rate of Profit $\frac{S - F}{C + V} = R$	Price of Production C'+V+F+P	Deviation from value to Prices
I_{R1}	80	20	50	20	90	25	95		107	+17
II_{R2}	65	35	55	35	125	5	95		107	-18
III_{R3}	70	30	60	30	120	15	105	12%	117	- 3
IV_{R4}	95	5	30	5	40	40	75		87	+47
V_{R5}	55	45	30	45	120	5	80		92	-28
VI_{R6}	60	40	35	40	115	10	85		97	-18
Total Social Capital	425	175	260	175	610	100	535		607	

acquires special importance in controlling prices and the cost of reproduction of the labour force. This is true not only in the well known relations between petty commodity production in agriculture and other sectors; of equal interest are modern cases of in-home-by-piece production in the cloth industry or in machine tools, and the sphere of domestic labour (Thompson, 1983).

Thus, the total mass of profit available for accumulation in a given region at a certain time will consist of two positive and two negative components of GTV. First, on the positive side we have profits, stemming from surplus-value or surplus labour extracted and realized locally (or elsewhere, but profits have been used for re-investment within the region), plus inter-class profit on circulation to the benefit of capital. Secondly, on the negative side, accumulation potential and corresponding profits will be eliminated due to direct and indirect GTV and/or due to losses of value. These joint effects <u>operate simultaneously</u> and could provide a framework for an analysis of sectoral and geographical differentiation.

Summing up, I have presented an investigation of uneven regional development in capitalist economies, arguing about its roots and the underlying tendency for its reproduction. In this respect, I have introduced first the notion of the contradiction of equalization/differentiation, and then the geographical transfer of value based on an established differential accumulation potential among the various regions of a social formation. GTV takes place in the sphere of circulation and exchange, hence its influence by the practices of civil society and state apparatuses; while its magnitude and direction (positive or negative) is determined in the sphere of production. As an analytical focus, the GTV is not aimed at submerging any other source of uneven regional development, especially in cases where political and cultural factors are historically important. Instead the GTV serves primarily to specify explicitly that uneven regional development in the CMP contains a set of powerful tendencies which produce and reproduce it. These tendencies are not automatic, they are dependent primarily on the operation of the capitalist economy itself <u>and</u> on the practices of specific social agents. In this respect, uneven regional development is not a simple and static reflection of the geographically uneven distribution of productive forces within a particular social formation. It is rather a dynamic process, acquiring various forms and intensities, depending on the specific historical conjuncture in the social formation under question. Capitalism requires uneven spatial development as a condition for profitable accumulation. But this unevenness may vary historically from region to region, within regions and among countries at the international scale. When we move from the abstraction of 'capital in general' to the 'concrete capitalism' of a social formation, space and territory prove to be a powerful means for differential accumulation potential, thus actively participating in the reproduction of social relations of production. Limits to the production of uneven regional organization - from a purely capitalist point of view - are therefore in a sense always temporary, because, while they emerge from the equalization tendency, they can also be reversed to restore profitable accumulation (e.g. the regional 'role reversals' of north England, Wales, Wallonia).

Thus, if GTV is influencing uneven regional development, it does so primarily as the outcome of a spatially-differentiated accumulation process. What this means in part is that GTV is the result of particular historical conjuncture rather than an originating process in itself. This should not be interpreted, however, to suggest that GTV (as an 'outcome' of uneven capitalist development) can be separated from the dynamic of capitalist development as a 'process'. Clearly the aim was to show how uneven geographical development both <u>shapes</u> and is <u>shaped</u> by GTV.

At a very fundamental level, the analysis will remain incomplete without a discussion of the state, its apparatuses for intervention and the wider ideological hegemony which prevent - at least temporarily - the endemic crises of the system. This discussion, which lies at the heart of the problem, i.e. the political implications of uneven regional development, will be presented for Italy, Spain and Greece at length in Part II, together with an analysis of regional social movements, which I see as the corresponding strategy for political struggle. In the next chapter, however, I will discuss certain wider theoretical and methodological issues concerning the state, local authorities and forms of struggle.

NOTES

[1] In Marx's words, '... it is evident that the balance among spheres of production of different composition must tend to equalize them with the spheres of average composition, be it exactly or only approximately the same as the social average.' (Kapital, Vol.III, p.173).

[2] Disarticulation and fragmentation simultaneously create an international division of labour which, as Marx argues, '... converts one part of the globe into a chiefly agricultural field of production, for supplying the other part, which remains a chiefly industrial field.' (Kapital, Vol.1, p.451).

[3] A. Markusen (1979b) argues in a classical aspatial economic framework that, '... capitalist production relations in their fundamental logic are spaceless...' and '... there is nothing in the logic of capitalist accumulation that requires spatial differentiation.'

[4] The idea is associated with the underconsumptionist view of capitalist crisis. While underconsumption is present at all stages of capitalist development, it becomes especially acute only in the monopoly stage of capitalism, see Sweezy (1956); Barran and Sweezy (1970).

[5] See Marx, Kapital, Vol.III, pp.198, 366, 644.

[6] See the summary of the debate made by S. Amin (1977).

[7] The debate between A. Emmanuel (1974) and B. Warren (1973), carried out in the pages of the New Left Review in 1973-74, clarifies the political significance of the theoretical conflict which the debate over unequal exchange brought into the open.

[8] See the criticism to regional applications of unequal exchange by Walker (1978), Markusen (1979a); Läpple and van Hoogstraten (1980), Webber (1982) and Cooke (1983). As I mentioned in Chapter 3, while I agree with most of the critics, I keep nevertheless the idea of surplus transfer as a potential explanation of uneven regional development, denying that unequal exchange is its prime channel. See also, De Janvry (1981); Marelli (1983).

[9] See Kidron (1974); Emmanuel (1969); Hymer (1972); Palloix (1972); Mandel (1975); Lipietz (1977).

[10] The Italians also found themselves in a direct conflict between international integration

and 'national' regional development. Their participation in the EEC has generated conflicts around the industrialization of the Mezzogiorno. In the early 1960s this led to a series of parliamentary crises, with motions not only from the Socialist and Communist opposition, but also from government party members. Italian industries, and especially FIAT, protested effectively on behalf of big business in the north that they could not simultaneously respond to integration at the international scale, after tariff abolition within the EEC, and organize major new plants in the south; see, Cottone (1972); Cinanni (1972).

[11] For a good general discussion on the controversy, see Elson (1979).

[12] See especially Kidron (1974) and Kay (1975).

[13] The regionalization used in this example is a hypothetical one, based on homogeneity and input-output flows. In reality, however, the smallest possible geographical aggregate will be used, for which statistical information concerning monetary expression of the value it creates can be obtained. That means the use of official statistical regional boundaries. From this initial regionalization a new one can be obtained, based on larger geographical summaries, using criteria of GTV flows. For an application see Chapter 9.

[14] Losses of value (waste in labour time) are simply one case when GTV does not occur. This could not lead, however, to universal statements as Walker's (1978), '... one must be cautious also about assuming that what takes place between unequal regions is a flow of value, when what is in fact happening is that labour is being squandered in the less productive region because it operates at less than average productivity' (p.29).

Chapter 5
THE STATE, LOCAL AUTHORITIES AND FORMS OF STRUGGLE

For the first time since World War II, the present crisis of capitalism appears, more and more, as a crisis of the state. Attention has been focused, in southern Europe and elsewhere, not just on the usual failure of the state to 'manage the economy', but on the need to restructure expenditures, policies and apparatuses. In this respect the 'regional crisis' has been directly associated with the crisis of the central state and the contradictions of local authority actions. In a period characterized in Europe on the one hand by the serious criticism of state intervention and on the other by the rise of regional social movements in many countries, the whole question of the limits and possibilities to state actions becomes crucial.

In this chapter I intend to complete the discussion of the contradiction of equalization/ differentiation and the geographical transfer of value through an analysis of the practices of the state and local authorities. Particular emphasis is also devoted to various forms of popular struggle around regional and local issues. The parts of the argument which are exposed here focus on the particular influence of state's intervention and popular struggles to the spheres of circulation, exchange and reproduction.

5.1 THE DEBATE ON THE ROLE AND FUNCTION OF THE CAPITALIST STATE

We have seen thus far how development theories of both the bourgeois and Marxist tradition approach uneven regional development. As I pointed out previously, one major inadequacy of both autonomous and transfer of surplus development theories was their limited concern with the capitalist state and local authorities, when the latter exist as part of an elected administration system. While the pluralist interpretation of the state-dominated Western social science until the 1960s (and still remains the ideological foundation of state legitimacy), a recent growing research interest in the state by Marxist scholars has turned the initial neglect to a race as to whose is going to win the greater attention.

There are, however, problems with the accounts of development which have been discussed, where the state appears as nothing more than a reserve power of the capitalist class, able to be mobilized whenever the class in power decides to do so. As Cooke (1982) correctly observed, the important constraints which are imposed upon state action by the existence of a class society and indeed the limitations upon what a dominant class can itself achieve are assumed away. Moreover - and here I am coming closer to my subject - the possibility of important variations in state actions and in regional planning in particular are only randomly studied. As an outcome, the relationship between the nation-state and its 'regions' is reduced to functionalist regionalizations (e.g. administrative regions, planning regions, etc.) while the regional implications of the various sectoral policies are never synthesized into a holistic socio-spatial approach.

It seems therefore that, while we are witnessing a growing interest in the state in general (with a useful distinction between the 'state in capitalism' versus 'the capitalist state', made by Clark and Dear, 1984), there is a lack of analysis of the state in particular, what Poulantzas (1978) calls the institutional materiality of the state in space and time. Such a contradiction seems to be analogous to the economistic interest with capital in general, with capital-in-particular, and its consequences for spatial organization, development and planning made a mere afterthought. We must consider with great reservations therefore arguments that seek to base their analyses of the capitalist state on solely economic and abstract foundations.

Four kinds of state theory can be identified in the light of the discussion on uneven regional development taking into account that a number of other specialized reviews are available: instrumentalist, capital logic, class-theories and neo-Gramscian theories of the state (see also, Clark and Dear, 1984; Cooke, 1983; Urry, 1981a).

In the instrumentalist theories, the state is viewed as a simple instrument in the hands of the capitalist class. It seeks to manipulate worker's resistance, capital demands, natural resources, etc., through specific policies (including regional policies) to the benefit of capital (Miliband, 1974; Toft Jensen, 1982). Miliband's thesis is that the state depends on the capitalist ruling class and acts as the willing instrument of that class. Close to this instrumentalist view is the state-monopoly capital thesis (Cherpakov, 1969; Boccara, 1971). In this, the modern capitalist state, following the evolution of some internal laws, has passed from the competitive to the big monopoly stage of the economy and is now totally dependent on monopoly (very often multinational) capital.[1] Central and local state policies are mainly oriented to secure high rates of profit for the monopoly sector both nationally and internationally (Lojkine, 1977; Damette, 1980).

Both of these approaches have been heavily criticized (Poulantzas, 1975; Clark and Dear, 1984; Holloway and Picciotto, 1978). The instrumentalist theory is deficient because there is no identifiable unified ruling class; the many subsectors of 'the' ruling class are divided over short-run issues and are therefore unable to attend consistently to the long-run reproduction of the system. Poulantzas (1975) rightly criticizes Miliband for neglecting the essential structural links between the bourgeoisie and the capitalist state. What makes the state a capitalist state, he argues, is not the class composition of the personnel of the state apparatuses but the position occupied by the state in the capitalist mode of production. The last point is also relevant for the inadequacies of state-monopoly capital thesis. The latter are criticized for (1) not specifying sufficiently the particular mechanisms through which monopolies impose their will upon the state; (2) viewing the state as an 'external element' to the economy, remaining neutral during the phase of competitive capitalism while intervening during monopoly capitalism, and (3) failing to deal with who decides on the appropriate strategy for intervention and the right time for its application.

Without following the simplicity of instrumentalist and state-monopoly capital theories, capital-logic theories view the state as a 'fictitious collective capitalist', the principal role of which is to respond to capitalistic interests (Alvater, 1973). Capital logic attempts to 'derive' the state from the category of capital and the essential elements of the capitalist mode of production. It thus includes the state in the totality of capitalist social relations and denies that this analysis can proceed from the separation of the economic and the political (Hirsch, 1981; Cooke, 1983).[2] Involved here is an attempt to see the state as a phenomenal form of social relations and to discover why economics and politics appear to be distinct under capitalism (Clark and Dear, 1981). Various efforts thereafter have been made to derive the form and functions of the capitalist state from the competition between capitals, from the need for regulation and from its responsibility for the material reproduction of the means of reproduction in order to secure social cohesiveness and to promote international expansion for national capital.

This problematic, however, despite its conceptual merit, relapses into a fairly traditional conception of capital as an abstract entity with an intrinsic logic based on economic categories, thereby missing the material specificity of the 'concrete' state (Poulantzas, 1978). These attempts, therefore, have ended in many cases in an ahistorical sterility, by reading social action through purely economic categories. They have been criticized for their inadequate theorization of the class character of the capitalist state and for their functionalist and reductionist vision of capitalist development (Urry, 1981a). As such, they require a holistic concept of civil society to overcome their tendency to reduce political relations and actions to economic relations and decision.

The idea that civil society has a specificity of its own which assists the structuration of political action in particular localities is first found in the work of Antonio Gramsci (1975). Gramsci, as we have seen in previous chapters, elaborated a new conception of hegemony and civil society and was able to break with permanent dualism and economism, expressing thereby an appropriate emphasis on the political. He has been criticized by the capital-logic theorists because he speaks of 'politics as an autonomous science'. In addition, they argue that, while he was against economistic identifications, he failed to provide any alternative analysis of the relation between the economic and the political (Holloway and Picciotto, 1978), However, this is an oversimplified statement, for Gramsci's emphasis on the political was not a simple shift of emphasis. It was a continuation of Lenin's argument in terms of the historical development of the capitalist state, as a non-functional, non-linear dialectical process. In other words, a process which is not determined by some inner logic of capital but which is subject to contradictions, breaks and leaps forward, all made possible through the political intervention of the revolutionary subject.

If we accept that the state is 'derived' from the category of capital and if by this we mean a non-linear development, then this could happen only when the derivation refers, above all, to the struggles of social classes. At the level of the state this means a political intervention, which was precisely Gramsci's prime interest.

These complexities are better theorized - though not fully adequately - in the so-called class-theories and neo-Gramscian theories of the state (Poulantzas, 1976, 1978; Dulong, 1976; Laclau, 1977; Buci-Glucksmann, 1975). A class theoretic analysis of the bourgeois state begins with an analysis of social classes and views the state as a locus of class struggle and a mediator between classes. This view has been mainly addressed by Poulantzas and has been used by others in urban and regional research (Castells and Godard, 1974; Preteceille, 1981; Dulong, 1976; Tsoukalas, 1975). The state is viewed as being relatively autonomous from the process of accumulation and from the relations of production. This approach overcomes the reductionism of instumentalist theories and facilitates an understanding of several crucial aspects of the capitalist state, among them the use of ideology to legitimize class rule; the existence of repressive apparatuses of the state; the resulting appearance of the state as a crucial object of class struggle; and the formation of class consciousness as a decisive subjective element in the historical development of class societies. The state, in Poulantzas' last work, is itself a relation and (more precisely) the condensation of class relations. It is neither an 'entity' on its own, nor totally part of modern capitalist society; it is a field of class struggle, a product and a determinant of contradictory social and spatial relations (Poulantzas, 1978).

A constructive criticism to Poulantzas' work came from Urry (1981a), who emphasized the inappropriate rejection by Poulantzas of the concept of civil society. Following Gramsci, Urry suggests that civil society is 'located' between the state and the economy. The state possesses a form which is given by its attempt to sustain the overall conditions under which profitable accumulation can take place within its national territory. In its effort to sustain the changing conditions for accumulation, the state in fact operates within the heterogeneous relations which comprise civil society. And Urry concludes that in its attempt to guarantee accumulation the state seeks to organize, legislate and orchestrate the diverse relations of civil society.

In the neo-Gramscian tradition, further emphasis is given to the role of ideological hegemony as a means of securing the cohesiveness of the capitalist system (Laclau and Mouffe, 1981). In these theories, the local authorities can become separate from the central state when it is not successfully reproducing capitalist hegemony. Dominant classes must secure the acceptance of the existing mode of exploitation by dominated classes through a continuous displacement - in time and space - of popular demands. These demands can produce a series of crises, ranging from the particular (such as legitimation crises, representation crises, regional crises) to the more general crisis of the state itself (Poulantzas, 1976).

These views and those related with 'crisis theories' (Habermas, 1975; Offe, 1974; O'Connor, 1973) have been criticized for their 'politicism' and 'structuralism'.[3] In short, class theoretic and neo-Gramscian approaches (with the exception of Urry, 1981a) can be faulted for a neglect of the economic. By assuming the 'displacement' of crises from the economic to the political and ideological sphere, these theorists generally fail to relate state crisis to the contradictory and uneven process of capitalist accumulation. This destroys, as Holloway and Picciotto (1978) argue, the unity of the Marxist vision by tending to locate the state <u>between</u> classes and <u>outside</u> the economy, rather than as one aspect of a social totality. According to the critics, a major inadequacy of this approach is that it starts with 'political' concepts, most notably with what they see as the central political category of class. This is supposed to be in sharp contrast with capital-logic, which tries to construct a specific theory of the political from the materialist categories developed by Marx in Kapital. But to counterpose these two approaches on this ground is to create a false polarity: the 'logic of capital' in Marxism is nothing but the expression of the basic form of class struggle in capitalist society. Each approach subsumes the other, but each is useful better to grasp the nature of the subject under study, thus complementing rather than excluding each other.[4]

Summing up, class theories of the state and especially the neo-Gramscian view provide us with an important distinction between the economic, the political and local civil society. It is precisely the difficulty of making the separation and at the same time the connection among these three categories that relates local forms of representation and government with capitalist development. As I discussed in the previous chapter, the unity and at the same time the separation between the spheres of production and circulation/exchange rests upon the nature of the commodity form which only releases the surplus value it contains when that value has been realized on the market. The nature of the relationship between labour and capital is one of exchange: surplus value is 'injected' by workers into the commodity at the point of production, while labour receives less than this in the form of wages.

The sphere of production relations *per se* is an area into which the state, either central or local, has the greatest difficulty in intervening. When it does so, it is always a reflection of class struggle, and usually takes the form of an indirect intervention, mainly via legislation; The state and the local state, however, do respond directly to pressure from capital and/or labour in the sphere of circulation and reproduction, of which exchange is a part. The clearest examples of this are those aspects of policy which can be broadly included under headings such as regional incentives to capital and labour, agency assistance, zoning, means of collective consumption, and physical infrastructure. Certain implications of these spatially-differentiated components of circulation and reproduction have been considered in the previous section. Here, it is necessary to emphasize that the growing state intervention in circulation and reproduction changes the whole process of indirect regulation performed by the LTV. This requires a modern conceptualization of the LTV to account for cases in which the state (through price regulation, norms of exchange, or special trade legislation) could weaken/strengthen the 'pure' LTV. While the commodity system remains anarchic - there is no explicit and direct allocation of social labour and the means of production - a certain social cohesion is required to ensure the reproduction of capitalist relations. This occurs through _a posteriori_ norms via LTV and through contradictory policies introduced by the state and the local authorities.

5.2 FROM NATION-STATE TO REGIONS: THE STATE AND UNEVEN REGIONAL DEVELOPMENT

The presentation in the previous section of the various views about the capitalist state has shown the important political differences which are behind each approach. These differences show clearly the wide range of materialist explanation available for the historical development of the capitalist state.

From a regional development point of view, these explanations can best be structured around what Poulantzas (1978) calls the material framework of spatiality and historicity, of territory and tradition. This is where the state acquires importance in relation to the historical process of uneven regional development. In fact, it goes back at least to the creation of the so-called 'internal or national market', based on the gradual elimination of 'internal' geographical boundaries for the free movement of capital, labour and commodities, the generalization of commodity exchange and the realization of value in the circulation of money (see Mandel, 1976; Lipietz, 1977). According to such conceptions, economic unity, as the essential element of the modern state, hinges upon unification of the internal market, and this in turn provides the basis for uneven regional development to occur. Thus, the state's specific materiality and its connection with uneven regional development is held to 'derive' from capital's needs for free circulation and establishment of generalized commodity exchange. The corresponding class analysis is deduced from the argument that both the nation and the modern state were created by commodity capital and that the formation of classes also takes place at the level of the nation and the state.

Such conceptions (which echo the

instrumentalist and capital-logic arguments of the previous section) form part of a dominant and tenacious Marxist tradition. They tell, however, only one part of the story and, if left, could easily end in reductionist explanations of the state and its relationship with uneven regional development. Three points are relevant for our discussion.

1. There is a general agreement that the very process of unification and modern state formation are at the roots of uneven regional development. This process, however, cannot be adequately explained by the generalization of commodity exchange and the free mobility of factors of production. As Poulantzas (1978) and Dulong (1978) argue, the need for unification of the so-called 'internal' market does nothing to explain why such unification is located at the level of the state. Furthermore, these conceptions tend to follow a linear explanation of unification/integration, where first valorization takes place in a few (supposedly autonomous) regions. Later, through unification, valorization becomes possible at the national level and finally during 'imperialism' at the international scale (Mandel, 1976; Palloix, 1975). But, as Wallerstein (1976), Kay (1975) and others have shown, such a linearity is practically non-existent due to early integration of the world capitalist system. 'Internal' accumulation is possible only vis-à-vis 'external' accumulation (e.g. long-distance trade), a condition which differentiated most European regions from the beginning of the industrial revolution.[5] Finally, what is it that defines the notions of 'internal' and 'external' from the point of view of capital? Or what is it that places a specific space 'inside', while other spaces are 'outside'? Such questions cannot be answered on purely economic grounds. As Gramsci has shown for Italy, for unification to take place the prior enclosure of precisely that space which had still to be unified was necessary.
2. From another point of view, emphasis only on commodity exchange and capital circulation can lead to functionalist and positivist conceptions of all the elements that are supposed to constitute the modern bourgeois

state and are also responsible for uneven regional development; territorially homogeneous and differentiated history, language, social classes and culture (Urry, 1981b; Giddens, 1981; Cooke, 1983). Again, the contribution made by Gramsci is particularly useful, especially those analyses related with the class structure of northern industrialized Italy, versus the agricultural south in the early 20th century.[6] In a certain sense these elements are often understood as transhistorical essences subordinated to the economic categories of accumulation (Mandel, 1976). I shall not enter here into the dispute over the exact nature of these elements and their relations with accumulation and uneven regional development. I will have the opportunity to develop my position further through the concrete analysis of southern European agriculture in Part II. Here I want to suggest that these elements are better understood as being part of civil society. Very broadly, the state attempts to establish and sustain hegemony over the contradictions inherent among these elements <u>within</u> civil society. This does not mean that hegemony entails social, ethnical, cultural or spatial homogeneity; it merely entails an articulation which affects concealment and deflection. Continuously new variations emerge in terms of changing class structure, the formation of new social groupings, the politicization of economic location, geographical restructuring and regional or local struggles.

3. From what I have said so far it is clear that I disagree with a number of contemporary writers who argue that the role of the nation-state is diminishing (Mandel, 1975; Amin, 1974; Murray, 1971). From this latter view it follows that the role of regions is also diminishing, or is unimportant to begin with (Urry, 1981b). The current internationalization of capital and the global market integration, however, do nothing to reduce the peculiar weight of the nation-state and the importance of regions (Warren, 1973; Massey, 1978b). This is so because changes in contemporary capitalist socio-spatial relations give to nation-state and to regions a quite novel significance. These changes include an increased differentiation between local/regional

capitalists and international capitalists; an increased importance of differentiated regional labour markets; an increased capital mobility to take advantage of all possible variations in profitability; an increased role of central state and local authority expenditure; an increased importance of regional differentiation in terms of the conditions for the reproduction of the labour force. These changes have been described by Palloix (1975) as the 'regionalization' of capital, homologous to the process of 'internationalization'. And when they are combined either with some kind of repressed local nationalism (e.g. Catalonia, the Basque country, Sicily, Corsica) or with the crisis of the state, they have contributed to regional social movements throughout Europe.

Thus the modern capitalist state and the regions become contradictory preconditions for the continued existence and the reproduction of the capitalist system at least in Europe. This is particularly evident at the level of the EEC, where a first and hesitant step towards integration immediately regenerated regional conflicts and competition.[7] It seems therefore more appropriate to look into the serious economic and political problems generated by national integration in terms of what Poulantzas (1978) calls 'deterritorialization': a process which follows the separation of the direct producer from his means of labour, and assimilates and homogenizes spatial segments by shifting their internal frontiers.

Deterritorialization could be seen as synonymous to delocalization of economic and political power. It is the process during which local variations in the practices of civil society tend to diminish. To put it in another way, the state realizes and secures its power through the unity of various 'private' agents and actors in a social formation. This is materialized at the very moment of their homogenization into a functional and deterritorialized entity.[8] The state needs a territory with specific frontiers to apply its power, both internally and externally. But the establishment of such conditions is not the simple effect of certain accumulation patterns, but the result of particular social struggles. To illustrate this, we could identify three forms of struggle: first, an internal class struggle among different regional capital fractions during early stages of the state's formation. These struggles revolved around the guidance and control of the monetary and taxation union and the development of the transport and communication system, necessary for exchange and circulation. Secondly, the struggle pursued mainly by popular forces to establish in all regions conditions of equality in commodity exchanges in terms of both price and availability. In this case we could include attempts to equalize the cost of reproduction of labour power across regions and sectors and the corresponding working class struggles in terms of wage, child labour, conditions of work and so forth. And thirdly, the struggles that extend the principle of exchange equality into the political arena. The combination, for example, of the constitutional rights, the right to vote, the right for self-government, etc. These struggles and the state apparatuses' actions tend to create an homogeneous political space where all people in all regions at least theoretically have equal rights. The latter in fact serves as a precondition for social and spatial differentiation and, in the last analysis, for the establishment of a more refined network of social and spatial exploitation. As Poulantzas (1978) pointed out:

> ... separation and division in order to unify; parcelling out in order to structure; atomization in order to encompass; segmentation in order to totalize; closeness in order to homogenize; and individualization in order to obliterate differences ... (p.107)

These were the roots for a centralized and deterritorialized state. The resulting geographical structure is a composite product of state action and other socio-spatial processes under capitalism. It seems, however, that state's role is increasing through the promotion of national spatial integration. The latter acquires constitutional status through the open negation of spatial diversity and isolation.[9]

The intervention of the state tries to submerge in the short term the negative effects of uneven regional development (which would undermine its territorial existence) using two levels of appearance. First, as the level of social aspirations and normative goals of society, the state attempts to establish its hegemony appropriate to the developing exchange relations. Secondly, at the level of action and conflict avoidance, the state is trying to establish conditions in which all individuals exchange commodities in a position of approximate legal equality. These appearances attempt to improve the conditions for the realization of value and compensate for the dysfunctional consequences of accumulation that have elicited politically effective reaction on the part of individual capitalists, organized labour and other social agents. The modern state *vis-à-vis* the economy and civil society assumes increasing influence mainly through the 'production' of regional space, a process that in previous phases of capitalist development was the outcome of the so-called free play of market forces. As Lefebvre (1974) argues, state intervention changed the 'conditions of production of commodities in space' to the 'production of space itself for the production of commodities'.

The various state apparatuses intervene directly in the production of regional space in a number of ways. O'Connor (1973) identifies three important aspects for state intervention: social investment to increase the productivity of labour; social consumption to lower the reproduction costs of labour power; and social expenses to maintain social cohesion. These three aspects are unevenly distributed over space, and the state attempts a double mediation. Technical mediation involves, for example, development of the transport and communications necessary for the 'displacement' and mobility of capital, labour and information. *Juridical mediation* encompasses property and land use laws, reorganization of spatial boundaries, redlining, etc. Technical and juridical mediation arise mainly through:

a. financing regional projects unprofitable for capital and by direct subsidies to capital to increase mobility, competitiveness, etc.;
b. designing and implementing specific regional projects, such as land use plans, new towns, growth poles, urban development and renewal, and national settlement policies;
c. intervention in the reproduction of labour force, through wage legislation, education and control over the unequal regional distribution of the means of collective consumption;
d. implementing sectoral policies (e.g. energy, transportation, tourist and agricultural policies) especially through the geographical distribution of funds and loans, which have an

immediate regional impact.

State apparatuses do not create the conditions of uneven regional development or the geographical transfer of value, but they contribute to them and so contribute to their reproduction and regulation. Avoiding an instrumentalist logic, this is so because the state attempts continuously first to sustain the overall conditions for profitable accumulation within its national territory; and secondly to sustain through regional planning and, whenever this is possible, conditions for profitable accumulation within each region. These two tasks are by definition contradictory. This is so because each state, in its attempt to sustain the changing conditions for profitable accumulation intra- and interregionally, in fact operates within the heterogeneous relations which comprise civil society and faces simultaneously national conflicts and local struggles. It is this contradictory location which Urry (1981a) suggests as central to our understanding of the state.

Let us now consider three aspects of civil society which are important for the analysis of geographical structure as a composite product of state action. First, there is the sphere of circulation and the activities of various private agents in exchange; secondly, the social relations within which labour-power is reproduced economically, biologically and culturally; and thirdly, social classes and other social groupings (Cooke, 1983). The state attempts to play a decisive role in controlling these aspects through the mediation of their extended reproduction. The latter involves a triple intervention that cannot exist in isolation from one another:

- First, there is an extended reproduction within the realm of civil society of the '_agents_' themselves; for example, small and big capitalist firms, petty-bourgeois commodity producers, unionized and non-unionized labour, the smallholding peasantry, 'feminished' work, etc.
- Secondly, there is an extended reproduction of the _places_ of these agents in the _social division of labour_; for example, the division between management, skilled and unskilled work or between the mental and manual labour.
- Thirdly, there is an extended reproduction of the _places_ of these agents in the _spatial_

> division of labour; for example, in core, semi-peripheral and peripheral regions; in uneven regional development; in segregated neighbourhoods in urban areas; in the dichotomy between urban and rural life; in the differentiation between place of living, place of work, and place of leisure.

The extended reproduction of social relations of production in these three aspects is not a contradiction-free process. With the appearance of functional and managerial weaknesses in the system, the generation of a number of side-effects from the intervention mechanism, and more importantly with the appearance and development of class struggle, the basic bourgeois ideology of fair production, exchange and regulation collapses. Thus the state and its apparatuses enter an open crisis, in which an important element is the reorganization of the spatial division of labour in the form of its core-periphery regional structure. This in turn directly introduces the regional crisis as an important element within the wider crisis of contemporary state.

State power is limited by two major institutions which it has itself created and maintained: private property in land and liberal forms of local representation (Poulantzas, 1976; Lojkine, 1977). Both are territorially constituted and administrated, and the state faces struggles arising from the attempt of different fractions of capital to establish specific spatial arrangements. Similarly at the level of popular and working class forces, struggles and social movements could arise to establish conditions of social equality at the local level and to extend this principle into the regional one to include issues of regional grievance, regional autonomy, etc. This very process of homogenization could generate conflicting territorial demands as a result of different local effects of the restructuring process. As Habermas (1973) argues, recoupling of the whole system creates an increased need for legitimation:

> The state apparatus no longer, as in liberal capitalism, merely secures the general conditions of production (in the sense of the prerequisites for the continued existence of the reproduction process), but it now actively engages in it. It must, therefore - like the pre-capitalist state - be legitimated, although

> it can no longer rely on residues of tradition that have been undermined and worn out during the development of capitalism. (Habermas [1973], Legitimation Crisis, p.35)

The need of the state to be legitimated lies at the heart of the problem of uneven regional development in Spain, Italy and Greece. As Gramsci has shown, when the bourgeoisie and the state apparatuses appear to face a legitimation crisis, they in fact face a problem with their 'ideological hegemony' over the social formation: they do not 'lead' the social formation, they simply 'dominate it'.

When the state and the ruling class are facing a crisis in their ideological hegemony, this means - among other things - that uneven regional development can no longer be considered legitimate. The various means of ideological domination - the news media, the educational system, the planning experts, the language, the church - cannot consolidate the image of uneven regional development as an inevitable outcome of material progress, or argue that 'regional problems' stem from the supposed inadequacies of particular regions and their people.

The legitimation problem is typically responded to with attempts to reduce the social formation to a system of structurally depoliticized public realms, spatially differentiated yet homogenized and fragmented. In this respect, civic privatism is promoted, that is a political apathy whose main emphasis is on the separation and fragmentation of career/security/recovery, from money/consumption/leisure, from place of work/place of living/place of recreation (Ledrut, 1975). Depoliticization, privatism and fragmentation require justification which is provided both by elite technocratic theories and by state planning.

The tendency, however, by the state to promote civic privatism and to reduce subjects to passive agents is neither always successful, nor could be taken as a conscious attempt. It is dependent on the articulation between the state and civil society and the development of popular democratic struggles. Thus, legitimation crisis could be a key source for regionalism and active citizens' participation when and where local regional objective conditions exist.

5.3 THE ROLE OF LOCAL AUTHORITIES AND FORMS OF STRUGGLE

I have argued so far that there is a need for an

analysis of the state in particular, especially in its spatial context. Extending the argument further, there is no reason for thinking that it is possible to develop a theory of the state in general, apart from particular kinds of society. So, it is incorrect to view the state functionally, either as reduced to the economy or as simply the state of the bourgeoisie. Given that there is a domain of civil society between the economy and the state, it is more constructive to analyse particular states in connection with particular forms of struggle and social conflicts:

> The role of local authorities and the nature of conflicts around local issues (at regional or community levels) must be seen in connection with these general trends. When I refer to 'local authorities', I assume a kind of elected self-administration system at different levels.[10] As in the case of the central state (and perhaps in a more direct way) the role of local authorities is determined by the historical evolution of local institutions. Therefore we cannot speak of local authorities in general, when important differences exist between north and central Europe, and southern Europe. In the former, local self-administration and strong local institutions date back to the Middle Ages, and the development of the central bourgeois state has institutionalized them. In the latter, however, while local experiences were there as well, the development of centralized and authoritarian institutions at the level of the central state destroyed most of the local initiatives (Tsoukalas, 1981). Elected local authorities in fact have been 'accepted' in Spain and Greece only after the 1970s, while regional council elections first took place in Italy in 1975.

We can begin noting that both liberal and certain Marxist theories for the local state in general can be accused of functionalism, since, by cataloguing action, they attempt to give a specific content. In this context, there is a tendency to reduce local authorities simply to an administrative apparatus of the central state. In some cases, as for example in certain municipalities in Greece, this could be true. We cannot, however, generalize the argument, since many forms and more than one type of local authority exist, depending on the articulation between central state and local political forces.

Thus, the debate on whether local authorities are autonomous from the central state and whether they are subject to direct manipulation by both capital and central state seems to be superfluous without a concrete historical analysis. It seems, however, that class theories and neo-Gramscian theories of the state permit the conception of local authorities with a relative autonomy as long as they have their own economic resources and are able, through local legislation, to control spending. From a Marxist perspective, this means to understand the role of local authorities in terms of the local social relations of production and reproduction (Duncan and Goodwin, 1982; Cooke, 1983). The interlinked nature of relations of production and reproduction develops certain local class relations, operated in particular local work places and community experiences, which, in their turn, give expression to the content of local authorities' actions.

We have to address here two major problems. First, the specific role of local authorities *vis-à-vis* central state apparatuses, given the relative autonomy of the former from the latter. Secondly, whether local authorities are the expression of local class relations, as opposed to the wider framework of national or even international class relations. Starting with the first, I think it is misleading to argue like Cockburn (1977) that local authorities take care of reproduction issues only, while central state orchestrates production at the national level. A similar view is adopted by Saunders (1981), who argues that local states' functions are mainly the provision of infrastructure, the organization of collective consumption for the reproduction of labour force and the maintenance of order through legitimation. These views, which correlate one set of social relations with one level of the state, while another appears elsewhere, are inadequate in explaining the role of local authorities. Local authorities do operate at a different level from the central state, but this cannot break the articulation between production and reproduction as it occurs at a particular time and place. Local authorities - to the extent that they have the power - intervene into the spheres of production, circulation and reproduction *parallel* to and not replacing the central state. For these reasons, local authorities are very often in conflict with the state and with one another. The problem is simply that the economy operates at a much wider spatial scale than is recognized by conventional political and regional boundaries. So, local authorities may have little impact upon, or control over, the process affecting local economies. This mismatch between political authority and the power to regulate the private market generates structural tensions between tiers of the state, the most significant being the recurrent call for reallocation of functions among various geographical levels of the state.

For the second problem, I am arguing (see also Paci, 1978; Bagnasco, 1980) that contemporary changes in labour processes and in the internal organization of the capitalist firm through Europe increase the importance of local class structure. I am therefore hesitant with analyses of class relations, when the basic unit of investigation is taken to be the nation state. The focus in these analyses is on male national classes and on the absolute and relative rates of mobility in and out of such classes. Yet, geographical variations and discontinuities in two important aspects, first in the _structure of regional labour markets_ and second in the _dominant forms of struggle_ could make the national class argument highly vulnerable (see also, Bleitrach and Chenu, 1979).

Starting with the first aspect, the structure of regional labour markets includes sectoral, occupational, cultural, ethnic and gender characteristics. The connection between labour market theories and regional development has been widely applied in Italy during the 1960s and 1970s (Magnaghi and Perelli, 1978; Garofoli, 1974; Arcangeli _et al_., 1980), and later in the neo-dualism thesis (Mingione, 1981). It has also attracted major interest among English-speaking scholars (Friedman, 1977; Urry, 1981b; Cook, 1983). The basic idea is that the development process produces, and in turn reproduces, a spatially-diverse division of labour. Some areas display a concentration of prestigious, well-paid jobs for which a substantial level of qualifications is a prerequisite; others tend to display the inverse of these qualities. Once established, this spatial division of labour changes over time, undergoing a permanent, dynamic process of recomposition in relation to the conflict between accumulation imperatives and worker resistance. The internationalization of capital and the growing intervention of central and local state create a fragmented local labour market able to provide the necessary stability and loyalty to corporate decisions (Friedman, 1977; Garofoli, 1983). Since labour power is produced and reproduced outside the factory, the relationship between capital and labour becomes one of double-dependence. It also creates a paradox: in an era of hypermobility and telecommunication, labour markets tend to be more and more decomposed and fragmented at the local level, while labour is willing less and less to move outside the locality for jobs.

Thus, territorial differentiation in the labour process and in the practices of civil society give rise to increasingly differentiated class practices. The latter increase the importance of local class structure (Urry, 1981b; Duncan and Goodwin, 1982). For example, an effective control over capital-labour conflict by political regional institutions could influence positively the development of the so-called regional market. If, for example, capital and labour in a specific region agree to a wage increase parallel to productivity, and the state manages to have an adequate supply of the means of collective consumption, then the value of labour power increases as well, and the working class will be better able to bargain for higher wages, which in turn may lead to an expansion of the regional market.

This fragmentation has profound spatial implications. The ability of many firms to survive in a crisis period is dependent, among other things, first on their capacity to take advantage of this spatially differentiated labour market framework; and secondly to shift locations of the main sites of production. A good example in this process is the productive decentralization in Italy since the 1960s and in Greece since the 1980s. It involves subcontracting parts of the production process to small, often family, firms located at a distance from the centre of strong unionization (Arcangeli *et al*., 1980; Bagnasco, 1980). The latter has the effect of keeping wages down, while avoiding strikes and other labour disputes. Similar restructuring - and perhaps more dramatic - is taking place in southern European agricultural and tourist sectors. I will develop these issues further in the next section. Meanwhile, I would like to notice the combination of seasonal native (mainly female) labour with 'foreign' labour (that is from other regions or other countries), which transformed deeply traditional social relations in the countryside. Thus, new and more highly differentiated regional labour markets come into existence because of the ways in which the production process is decomposed. This works to the advantage of capital and the central state, since it fragments the interregional solidarity of working class which is organized on a functional, rather than a territorial, base.

The second aspect, the dominant forms of struggle, includes a variety of practices of civil society, from class struggle at the point of production to struggles around cultural, ecological, gender, regional or urban issues. Here particularly helpful is the distinction made by Laclau (1977) between 'class struggle' and 'classes-in-struggle'. In the former there are capitalists and workers who are in conflict around the production and appropriation of surplus value. Neither class can exist except in a relationship of struggle. This relationship, however, is not only an economic one. Classes are not abstract categories: as Touraine (1978) shows, they exist in struggle only within civil society. For Touraine, society is not analysed as a social system driven by an inner logic, but as a field of social action. This means that the forms of class struggle are given by certain characteristics which include both general accumulation patterns (such as sectoral, juridical, national, international) and particular-to-a-place conditions (such as the realm of local civil society, the political organizations of local authorities, the structure of regional labour markets, etc.).

'Classes-in-struggle' are those categories of subjects who occupy a common position in relation to the means of production but do not entail direct antagonism. Examples of such classes would include the new middle class, the state bourgeoisie, the traditional petty bourgeoisie, and the lumpen proletariat. In the case of classes-in-struggle, their form and effectiveness are more directly dependent on the nature of the local civil society and its articulation with the state and local authorities (see also, Cooke, 1983). This is why the spatial element in these forms of struggle acquired growing importance. For example, the central state versus a local authority; a large multinational versus the local people; EEC regulations versus regional demands, etc. (Hadjimichalis and Vaiou-Hadjimichali, 1979).

Another dominant form of struggle is that of

'social movements' (see Touraine, 1978; Pickvance, 1972; Castells, 1983). Social movements account for the ongoing creativity of social actors in developing a plurality of new forms of democracy such as councils, local assemblies, democratic associations, and so forth. For Touraine (1978) the locus of action for social movements is civil society. He identifies a social movement as:

> ... the organized collective action in which a class-actor struggles for the social definition of historicity in a given historical ensemble. (Touraine, 1978, p.35)

Thus, class struggle and social movements become synonymous expressions for the contestations involved in the control over 'historicity'. But historicity has no meaning without a concrete spatial reference. It is therefore surprising why Touraine fails to identify the 'spatiality' of capitalism (see particularly, Soja, 1980), which in fact gives a 'place' to the 'given historical ensemble'. Social movements take many forms, as the expansion of the arena of social contestation includes extra-economic domains such as ecology, feminism, peace, regional and urban issues. These developments are accompanied by an increased reflexivity regarding the social construction of reality and social identity as a whole.

For example, a thorough analysis of urban social movements and their relationships with central state and local authorities is provided by Castells (1983). He is making a complete departure from his earlier model, de-emphasizing now his theory of collective consumption and giving attention to Touraine's theory of social action.[11] The social movement operates on three fronts: collective consumption, cultural identity and political self-management. It makes use - but it must remain autonomous - of the media, professionals, the state, local authorities and political parties. Finally, Laclau (1977), using a similar approach to that of Touraine in describing social movement, suggests that there are 'popular-democratic struggles' involving the organization of the 'people' based on non-class forms of interpellation. This is very important since it means, as Urry (1981a) depicted, that class struggles and classes-in-struggle take place within an already structured civil society, structured in terms of the particular effects (in time and place) of popular-democratic struggles.

From the presenting discussion three points could be made. First, contemporary changes in regional labour markets and local forms of struggle throughout Europe increase the importance of local class structures within the realm of civil society. Secondly, local class relationships - as well as other forms of social relations - give expression to the content of local authority's action. The latter may or may not favour certain capital interests or central state's policies. In those cases an antithesis could arise with profound political implications. In a wider political context the content of central state's policies (e.g. regional policy) is also an expression of class antagonisms at the level of the state and struggles for political domination at the level of civil society. Thirdly, local authority's actions and central state's policies - especially in the spheres of circulation and reproduction - could make a certain region more-less attractive for investment and more-less competitive for surplus value realization. In those cases a potential for differentially localized accumulation is set up, which may give rise to geographical transfer of value. The net effect of these trends is a forced and increased 'politicization' of the spatial structure of capitalist social relations. The tensions between central-local state, together with the continuous capital restructuring and the reallocation of political functions among the various tiers of government have the effect of regionalizing the crises of contemporary capitalism.

NOTES

[1] This is the official explanation used by PCF and KKE in their analysis of contemporary France and Greece. Both parties have used the state-monopoly capital thesis to oppose EEC. See also, the interesting debate between Poulantzas and Lojkine summarized by Pickvance (1972) and Lebas (1978). Lebas made a mistake, however, arguing that the debate was between two members of the same party. Poulantzas was never a PCF member; he was, until his death, a member of the KKE of Interior.

[2] This problematic has been developed, particularly in West Germany, under the name _Ableitung_, and in Great Britain and the United States under that of _Derivation_. See, Holloway and Picciotto (1978).

[3] The Althusserian influence on those theorists was severe. Some of them (as Poulantzas, Castells, Dulong and others) have overpassed this influence in later works. See, for example, Dulong (1978).

[4] It is thus unfair and incorrect to put together Miliband and Poulantzas - as Holloway and Picciotto (1978) and Cooke (1983) do - in order to present capital-logic as the most appropriate explanation of the current function of the capitalist state.

[5] Pierre Vilar (1962) in his nominal work on Catalonia shows the interrelationship between 'internal' and 'external' factors that differentiated Catalonia from other Spanish regions. He uses a threefold dialectic (a) between 'long times' and specific times of the mode of production, (b) between the small spaces of ethnic groups and the large zones of 'external' modern trade activity, and (c) between class struggles and the consciousness of local groups.

[6] See, Gramsci's notes on 'The Function of Piedmont', the 'City-countryside Relationship' and 'Political Parties in Periods of Organic Crisis', in Prison Notebooks.

[7] The first meeting of European regions organized by the EEC in Marseilles in 1984 has shown the deep differences among them, and at the same time the potential for common action.

[8] See also, R. Ledrut (1975) and Claude de Vos (1975) in the IXme Colloque de l'Association Internationale de Sociologues de Langue Francaise.

[9] Interesting examples are the Greek Constitution of 1975 and the Spanish Constitution of 1978, both introducing state's obligation for integrated development. The Spanish one goes further to acknowledge political regional autonomy (Articles 2, 143, 151).

[10] I refer here to local authorities instead of local state to make the distinction of self-government traditons between north-western and southern Europe.

[11] For a presentation of the 'old' Castells' model, see Castells (1972).

Part II
EMPIRICAL INVESTIGATIONS – AGRICULTURE AND UNEVEN REGIONAL DEVELOPMENT IN SOUTHERN EUROPE

The discussion of the political economy of uneven regional development, the underlying contradiction of equalization/differentiation and the 'produced' tendency of the geographical transfer of value, provided the analytical tools for a critical understanding of socio-spatial unevenness in southern Europe. A primary conclusion from the discussion in Part I is that the development of certain regions and sectors within a social formation is not independent of the underdevelopment of others. Once a region has moved ahead (taking advantage of historically established 'internal' and 'external' conditions), this region may start to make others pay for its prosperity through direct and indirect GTV. This process is dependent on struggle (among capitalists, between labour and capital, among localized interests, etc.) and the political intervention of collective actors, usually the state, local authorities and social movements.

Thus, a regionalization tends to be established within every social formation (and within the world capitalist system) based on a hierarchical social and spatial division of labour. To define this regionalization I use a twofold approach (see also, Lipietz, 1977; Massey, 1978a). First, a regionalization defined in terms of regions' histories, that is the totality of social, political and cultural relations which they experienced in the past. And secondly, a regionalization defined in terms of present regions' relations, based on surplus transfer to the currently emerging spatial division of labour.

This regionalization, once established, changes over time, following on the one hand the new requirements of capitalist restructuring and accumulation, and on the other the actions of different classes and social agents. This means that an analysis along these lines could not restrict itself to purely economic categories to explain regional unevenness. GTV is an underlying tendency, but its explanatory power must include an analysis of social classes in particular regions and localities, to show how social action creates the conditions that make GTV possible.

In so doing, I propose to approach the problem of uneven regional development in Spain, Italy and Greece through an analysis of their agricultural sector for four main reasons. First, agriculture has been a neglected sector in regional analysis, yet its economic, social and political importance during the long post-war boom and the rapid industrialization period from the early 1950s to the mid 1970s remained at high levels. Secondly, in these early stages of rapid industrialization in southern Europe, agriculture was 'squeezed' to support the overall rate of growth, and this relationship was determined critically by the size of the market surplus and the term of its transfer. My ideas about GTV originated in the empirical analysis of various forms of agricultural enterprises, their relationships with other sectors and their regionally-specific characteristics. Thirdly, intersectoral transfers from agriculture to other sectors and regions have been manipulated to promote particular class interests. This was significant to the class configurations and political alliances at the level of the state, which acquired a key role to secure the surplus transfer. And fourthly, these developments have an important historical dimension building upon early regional differentiations. The long history of changes in the southern European spatial division of labour highlights the importance of agricultural regional specialization and its effects on rural struggles and regional social movements. In this respect, the empirical investigation that follows is not aimed at 'proving' the theory described earlier. It rather serves as a starting point to show possible directions for further research.

Chapter 6
THE EMERGENCE OF REGIONAL DIFFERENTIATION IN SOUTHERN EUROPE

Uneven regional development in southern Europe is not a 20th century phenomenon. Regional problems were in fact consolidated by the beginning of this century, but their origins must be traced back to the 16th century (Braudel, 1972; Wallerstein, 1974). It was during the 'long sixteenth century', 1500-1640, as Braudel (1972) calls it, that an international social and spatial division of labour started to be established through a series of historical transformations.

Describing the process of capitalist development in southern Europe in the course of four centuries would be an enormous task, if it were to be treated thoroughly. But in the chapters that follow the analysis centres around a number of themes that I consider most directly relevant for the discussion of current regional problems. These themes are the pattern of accumulation, various forms of struggle, interregional and international relations and exchanges, the role of the peasantry and the rising importance of the indigenous bourgeoisie and the state. To avoid the danger of overly generalized schematization, these themes are examined through the concrete examples of Spain, Italy and Greece, with supporting reference to the wider Mediterranean and European context.

6.1 EARLY REGIONAL DIFFERENTIATION: 16TH AND 17TH CENTURIES

During the sixteenth century the main economic activity was agriculture, and the Mediterranean was a world of peasants, landlords and merchants. Southern Europe was able to live largely by its own agricultural production. No pattern was to emerge here comparable to that developing in the Lowlands or in England in terms of agricultural surplus (Braudel, 1972). This general specialization in agriculture was helped by the very moderate climate. Everywhere in southern Europe the same trinity can be found: wheat, olives and vines, born of the climate and tradition and produced in most of the regions bordering the sea.[1] The Mediterranean regions, therefore, were then in competition with each other for markets for their common agricultural products, as they are today within the EEC. Small initial differences in regional specialization and productivity developed into enormous regional disparities. With reference to the growing disparity between western and south-eastern Europe, Stoianovich (1971) points out that:

> ... if in the fourteenth century, one discovers little quantitative difference between the iron orientation of Balkan societies and (western) Europe's iron orientation, that distinction was significant in 1700, much greater in 1800 and incredibly greater in 1850. (Stoianovich, 1971, p.223)

The western part of southern Europe was subdivided into a large number of political entities (city-states, kingdoms, etc.) by the 16th century. It specialized in high quality agro-products (wine, olive oil, fruits), the provision of credit, and commercial transactions (Wallerstein, 1974). From the Atlantic coast of Portugal to the Adriatic coast of Italy, a complicated system developed, involving large-scale land ownership and family holdings, mortgages and rent with numerous entailed properties. In these properties wage workers and serfs were struggling for life: according to moderate estimates, 20% of the population were living in extreme poverty and were ready to migrate to the cities, where they often became vagrants and vagabonds (Lopez, 1973).

The population density was basically much higher in the western part of southern Europe, even in periods of demographic decline. As a result, agriculture was more intensive here than in the eastern Mediterranean. In the Iberian and Italian peninsulas and in the Languedoc region of southern France, the mode of labour control was based primarily on share-cropping (Wallerstein, 1974). Share-cropping was an intermediate form of labour control in agriculture, between free labour (applied mainly in northern Europe) and forced and/or coerced cash-crop labour (applied mainly in the eastern Mediterranean and northern Africa). This intermediate classification of agricultural labour permits Wallerstein to characterize the western part of southern Europe as a semi-periphery during the 16th and 17th centuries.

Land was generally controlled by the nobility and the Catholic church and fell under one of the two basic land structures: latifundia, large-scale holdings, and minifundia, small-scale holdings. These two forms of land ownership played a key role in the uneven regional development of Spanish and Italian capitalism: minifundia in northern Spain and Italy, and latifundia in the south of these two countries provided the basis for regional differentiation.

According to Malefakis (1970), land ownership differentiation between northern and southern Spain originated during the Reconquista against the Moors (c.1250). This differentiation was the result of a complex interaction among four factors: royal power, the nobility, the size of the districts conquered in different periods, and the method of repopulation of the districts. The distinguishing feature of southern reorganization was a repopulation via military orders, rather than by municipal councils, of 'presura' (free peasants competing for land parcels). The military order and big aristocrats permitted whole villages to be founded by charter as serf-villages, in which peasants laboured as serfs for the landlord, who owned both the land and the village.

Land ownership differentiation between northern and southern Italy had its roots in the Roman times (Cottone, 1972). In the south, the Roman conquest, mainly an enterprise of the bourgeoisie and the Senate, reflected the views and interests of the victorious upper strata. Large estates based on a Roman variation of absentee landlordism thrived on exploitation solely for profit. During the 16th and 17th centuries this 'latifondo contadino', dominated the entire south. Thus, most of the latifundia were worked by exploited serfs who did not have any direct interest in the land, which gradually became less and less productive (Barzanti, 1965). Centuries of these relations, aggravated by climatic conditions, scarcity of water and poverty of soil, brought about sharp differences between the agricultural conditions in northern and southern Italy.

The eastern part of southern Europe underwent

a process of development that was altogether different from that of the west. Development in south-eastern Europe was largely defined by Ottoman occupation, under which the area remained until about the middle of the 19th century. Ottoman administration controlled the spatial organization of annexed regions through a sophisticated system of land ownership accompanied by appropriate legislation and, when needed, use of direct force. This determined to a great extent population and production distribution and settlement structure throughout the Empire, following centrally defined goals.

Land was generally owned by the state but fell under one of the following categories, corresponding to different modes of labour control:

a. 'miri' land (domain belonging to the state) where peasants worked without having the right to migrate. The *raaya* (peasants) could move from one region to the other only under official (forced) population transfers (Moskof, 1974).
b. 'timar' land (domain owned by the state but rented to rich Greek or Turkish officials). The *raaya* were here more flexible but still under the absolute power of the *timar* leader. With the decline of the power of central authority, the largest part of *timar* land was gradually transformed by the 18th century to a system of latifundia, the so-called *ciftlics* (Mouzelis, 1978).
c. 'vakouf' land (land owned by religious institutions). Islamic law prohibited *vakouf* land from becoming *miri* or *timar*, and Ottoman presence there was limited to tax-collection. Peasants living and cultivating this land were considered 'free' and were under the protection of the religious institutions (Svoronos, 1972).

These land ownership categories in Greece produced a distinct spatial organization at the regional scale but also at the scale of each settlement and even each individual house.

When forced population transfers decreased, peasant families found themselves working on the same piece of land generation after generation. Even though they never officially came to own that land, they acquired it *de facto* - a situation that Greece and the rest of the Balkan countries had to cope with after independence. During the Ottoman occupation, regardless of land ownership relations, the raaya did not pay rent in money, crops or extra labour. Local leaders exercised control through a sophisticated system of taxation. This was a basic difference in the mode of labour control from that of their counterparts in the west.[2]

Western type labour control mechanisms did exist, however, especially in Greece, where until the 17th century many of the Aegean Islands, part of the Peloponnese and Crete were under the domination or influence of Venice and Genoa. In these regions, private land ownership was still combined with serfdom. Particularly in Crete there existed what Wallerstein (1974) calls 'coerced cash-crop labour', a system of agricultural labour control wherein peasants were required to work part of the time on a large domain producing for the European world market.

From the 15th century onwards, administration in the Ottoman occupied areas was organized on a regional taxation base. The whole empire was divided into '<u>ziamet</u>' (regions) corresponding to military fiefs and self-sufficient agricultural areas. Every region was subdivided further into <u>timar</u> or <u>miri</u> or <u>vakouf</u> prefectures, according to local circumstances; and every prefecture was focused around clustered settlements, villages and towns (Tsoukalas, 1981).

Synthesizing the comparative economic and social conditions that corresponded to different regional structures in the western and eastern parts of southern Europe through the 16th and 17th centuries, the following four observations can be made:

First, in the east an almost stable population distribution was imposed by Ottoman rule, which thus avoided problems of unregulated migration to cities but at the same time eliminated communications and innovation diffusion among regional units. The most important characteristic of south-western Europe at that time was the opposite: high mobility and extensive exchange of ideas, people and commodities that gave a special role to large and medium size cities.

A second observation refers to the special role of cities and the urban based bourgeoisie in the west and the specific role of small agricultural communities in the east. Over-concentration of population and wealth in the capital of the Empire and relative autonomy of small villages did not permit regional capitals, such as Sofia, Thessaloniki, Izmir, Bursa, Ankara and others to flourish like the major urban centres in the west. Braudel (1972) argues that the weakness of towns in the eastern Mediterranean is an important element which differentiates development paths between east and west. This was a small difference in the 13th century but developed into a pronounced gap in the 16th.

A third major difference which derives in part from the previous two concerns regional productive specialization and external relations with the emerging world system. While in the west each region was participating in the world market on its own, in the east the central authority decided which crops had to be produced, in what quantity and where. In this respect it was the Empire which was the main economic factor, rather than the regional units, as was the case in Italy, Spain and Portugal at that time. During the seventeenth century, when the central power started to decline, regions in Greece took the opportunity to participate with greater economic autonomy in the international market. As in the case of Thessaly and Macedonia, the mountainous communities started to produce large quantities of textiles for export to the West.[3]

Finally, there was a sharp difference between the modes of regional accumulation. The entire social and political structure in the East was so organized as to facilitate collection of surplus and its transfer to the Empire's capital. The lack of private land ownership, the taxation system and the relation with the world economy via the central authority reduced regional accumulation to minimum levels. This was a basic difference from the west, where attempts for regional accumulation organized by a regional bourgeoisie were the prerequisite for development of commercial capitalism. What was occurring was a transfer of part of surplus being produced from one zone to another, forming a series of core-periphery relationships. In the east, there was one dominant core based in Constantinople. In the west, however, political and economic heterogeneity permitted many cores to flourish. As Wallerstein (1983) argues for the world capitalist system:

> ... as this process began, the spatial differentials were rather small, and the degree of spatial specialization limited. Within the capitalist system, however, whatever differentials existed ... were exaggerated,

> reinforced and encrusted. What was crucial, in this process, was the intrusion of force into the determination of price ... Unequal exchange is an ancient practice. What was remarkable about capitalism ... was the way in which this unequal exchange could be hidden. (Wallerstein, [1983], pp.30-1).

6.2 FROM CORE TO PERIPHERY: TRANSFORMATIONS OF INTERNAL SOCIAL AND SPATIAL STRUCTURES IN SOUTHERN EUROPE

During the 18th century, the world system underwent a reorganization of its core-periphery structure. The western part of southern Europe, which had been a core area during the 16th century, gradually became peripheral to rapidly industrializing north-western Europe. In the eastern part, the regions of the Ottoman empire and, later in the 19th century, the independent states, became part of the world system as peripheral units. It was north-western Europe that became the core of the world system, as the latter changed its social and geographical division of labour. As Lopez (1973) describes it, strong state apparatuses in the core were capable of supporting effective industrialization through a combination of four characteristics that southern Europe lacked: (a) bureaucratization, (b) monopolization of force and technological innovations by certain regions, (c) creation of legitimacy, and (d) homogeneity of subject populations (see also, Vilar, 1962).

For the new core area, however, industrialization involved substantial diversification of agricultural activities. In the 18th century, England was not only Europe's leading industrial exporter but also one of Europe's leading agricultural exporters (Lopez, 1973). In contrast, the land tenure system, the production process and the mode of labour control in Italy and Spain remained in the 18th century much the same as in the past (Vicens, 1969). A considerable proportion of arable land was owned by the Catholic Church and was left largely unexploited. Operating in close collaboration with the aristocracy, the Church legimitized the mechanisms of peasant exploitation and suppression.[4] The standard of living and the purchasing power of the masses declined significantly compared with those of the north. A purchasing power ratio of 1:8 was estimated between southern Italy and England at the beginning of the 18th century

(Braudel, 1972).

In Italy, regional accumulation continued to be the leading pattern. Internal markets, though, were not large enough to survive the competition of north-western industrial capitalism, and northern Italy saw its early industries decline rapidly. In Spain, on the other hand, exploitation of the colonies was the main mode of accumulation, even though central power declined. Investments, however, were concentrated in unproductive activities. Thus, by the 19th century, when most Spanish colonies gained independence, Spain itself had become a depressed agricultural region with only a limited internal market (Vicens, 1969).

In the eastern part of southern Europe, the Ottoman Empire participated in a process of commercial exploitation that many called semi-colonialism. After the 1740 capitulations and especially after the 1838 Anglo-Ottoman Trade Agreement, the Empire was forcibly opened for foreign trade. In the case of Greece, the Anglo-Ottoman Agreement resulted in the decline of the still weak manufacturing structure in mountainous villages but contributed to the rapid development of nagivation in port cities and in the islands, whose growth was enhanced by increasing opportunities for commercial activity. Like many other regions of the Empire (e.g. Anatolia), Greek mainland regions lost both productive activities and the regional specialization achieved in the 16th century (Svoronos, 1972).

At the end of the 18th century, Greece was governed mainly through the channels of local authority. Since central power declined, more capital could be accumulated regionally and more surplus could remain where it was produced. Formerly state-owned land was appropriated by regional leaders and turned later to *ciftliks*, i.e. large land properties, thus violating Ottoman law. But the Empire was neither able nor willing to suppress these leaders and still regarded them as its representatives (Moskof, 1974). From 1828 (the official end of the War of Independence), there was an attempt by the newly-formed Greek state to deal with *ciftlik* owners and, at the same time, with the peasant families who had worked on the same piece of land for generations and who now demanded official recognition of their property rights. The German-Roman Law became the key to manipulating these complicated property problems.

This attempt aimed at the direct introduction

of capitalist relations in agriculture, or 'social integration of agriculture', as Vergopoulos (1975) calls it. This was made possible by the early land redistribution and agrarian reform movement of 1871.[5] At that time, the Greek state owned about 35% of arable land, which was distributed to landless peasants. Large properties remained untouched, however, especially in Thessaly, and specialized in cereal production. In the south and in the islands commercial crops for export (sultanina raisins, vines, olives, figs, etc.) were the predominant mode of agriculture, favouring small family holdings and establishing what remains until today a primarily peasant form of production. Specialization resulted in a territorial antithesis: large properties in north and central Greece versus small properties in the south, a situation which had a variety of political and social implications (Mouzelis, 1978).

Looking at the overall development of the social formations in southern Europe, three major phases of 'semi-peripheral' and 'peripheral' development (from the 16th to the 19th centuries) stand out, directly related to corresponding phases in the ever changing international division of labour:

a. The first phase occurs in the 16th and 17th centuries when Italy and Spain gradually moved from core areas of the world economy to semi-peripheral positions. The Greek territories of the Ottoman Empire experienced indirectly, as peripheral units, the impact of rising north-western capitalism and the creation of a relatively coherent world market. In both the eastern and the western part of southern Europe the impact of these changes in the regional structure became evident later in the eighteenth and at the beginning of the nineteenth century.
b. The shaping and integration of southern European countries in accord with the developmental needs of north-western Europe assumed a more drastic form during the second half of the 18th century. Industrialization in England and the continent, and war and colonialism in Africa and the Americas destroyed the markets of Spain, Portugal and the industries of the Piedmont Valley. In Greece the destruction of handicraft industries put a full stop to the feeble attempts at an

endogenous industrial 'take-off' (Mouzelis, 1978). The spatial distribution of these effects resulted in the deterioration of the conditions in backward agricultural regions and established the 'North-South' dichotomy, not only between north-western and southern Europe but within southern Europe as well.

c. Finally, towards the third quarter of the 19th century, north-western imperialism took yet another and more aggressive form vis-â-vis southern Europe and especialy Greece. In the previous periods, it had made the integration of indigenous capital into the international market almost impossible. Now, north-western capital, in the form of railway investments and loans to governments, penetrated southern economies and contributed to considerable transformation in the social and spatial structure.

6.3 THE DEVELOPMENT OF INDUSTRIAL CAPITALISM: BOURGEOIS CONSOLIDATION AND POPULAR EXCLUSION

The end of the 19th century and the beginning of the 20th mark the stabilization of the conditions of accumulation in north-western Europe. At a time when Spain and Italy were still struggling for unification and Greece for annexation of national regions still under Ottoman occupation, north-western European states enjoyed a political and economic security that permitted them to dominate the southern countries politically and economically more than ever before.

In previous centuries, southern Europe had enjoyed a certain degree of economic and political autonomy that was associated in part with regional differentiation. From the middle of the 19th century, unification efforts in Italy headed by 'northern' leaders faced strong regional opposition (Nocifora, 1978). Unification was the demand of industrial capital, since it would favour the rapid development of national markets. In Spain these issues became important during the debate among bourgeois leaders in terms of a cultural and economic unification. In both Spain and Italy the process towards the formation of a national market was guided by a specific regional bourgeoisie, whose role was not acceptable to the other regional bourgeoisies. In Italy, the northern industrialists of Piedmont, Lombardy and Veneto assumed this role; in Spain the northern Catalonian and Biscayian bourgeoisies were predominant.[6]

Following different historical origins, the leadership for the industrialization of Greece came from abroad, from Greeks involved in the Diaspora. These were rich merchants and industrialists, who, until then, had lived along the Black Sea, in Constantinople, Izmir, Alexandria, Cairo and Venice (Tsoukalas, 1975). In this respect, no such conflict emerged in Greece among regional bourgeoisies and no region seemed at that time to have enough power to dominate the others. This was an important reason for the strength of the Greek state from its early formation and for the growing concentration of political and economic power in the national capital.[7]

Italy

The major regional problem in Italy was the already considerable gap between north and south. At the time of 'Risorgimento' (unification), southern Italy was far behind the north on almost any variable. For example, in 1861 the Kingdom of the Two Sicilies could boast of only 98km. of railroads, while in the same year Piedmont had 803, Lombardy 202, Veneto 298 and Tuscany 256. Furthermore, since 1867 the growing industry of the north was favoured by protectionist policies which kept foreign capital out and secured dominance over the domestic market. As Gramsci argues:

> This protectionism provided the basis of an effective commmunity of interests between big industrial capital and the reformist working class organization. (Gramsci, 1975, p.87)

This 'community of interests' permitted Cavour and later Giolliti to strengthen the dominant position of the north through the Risorgimento. In 1871 Cavour extended the Piedmontese tariff system to the whole country. This act caused considerable economic stress and was particularly damaging to southern Italy, which saw its traditional high tariff protection wall reduced by 80%. The high protectionism before the Risorgimento in the south permitted, according to Donolo (1972) a certain degree of industrialization. The impact of removing the tariff protection from the south was to cause a collapse of existing industry there and confine industrial accumulation to a minimum (see also, Myrdal, 1957).

The cultural and economic stagnation of the

south was further intensified by a process of direct geographical transfer of value: draining away capital from Mezzogiorno, carried out by the Public Treasury of the new state and by private financial institutions. A full process of 'de-industrialization' of the south to the benefit of the north occurred in this period (Nocifora, 1978). This became evident during 1880 with the programme of Giolliti and the democratic liberals who had the aim of creating an 'urban' bloc of industrialists and workers in the north, reinforcing the economy and northern hegemony. The Mezzogiorno was reduced, according to Gramsci, to a 'semi-colonial market, a source of savings and taxes'.[8]

With the exception of the central and northern cereal-producing regions, the impact of northern indigenous industrialization on agricultural Italy was undoubtedly very severe. Peasants were no longer able to export their products freely, due to restrictions and regulations imposed by the new central state; yet they were nevertheless forced to buy the products of northern industry and not the cheaper goods coming from northern Europe. Thus, 'southernism' - a type of peasant socialism - emerged as a reaction to this situation and spread in the south. Southern peasants were organized and struggled on a territorial basis following liberal and anarchist ideas and were not in line with the PSI (Socialist Party of Italy), the trade unions, or later with the PCI (Communist Party of Italy). Until 1922 these peasant struggles and a growing tendency towards radicalization of wider population groups provoked many cases of armed intervention.[9]

Between 1916 and 1921 the militancy of the peasantry gave rise to separatist regional movements. In Sardinia a strong regional separatist movement developed, the so-called 'Sardism', which split in 1921, one section joining the Fascists, while the other joined the opposition (Novacco, 1976). Similarly, a Sicilian Reformist Party appeared, whose extreme separatist wing was represented by 'Sicilia Nuova'. In the Mezzogiorno the group 'Rinnovamento', made up of war veterans, unsuccessfully attempted to start regional action similar to that in Sardinia. In these separatist movements, however, the autonomous importance of the peasant masses decreased progressively over time, as these movements came to depend primarily on the organized strength and the ideological pressure exercised by local big landowners and the rural petty bourgeoisie.

The peasant struggles in the south and the necessity to organize an alliance between peasants and urban proletariat was pointed out early on by Gramsci. He was the first socialist leader in southern Europe to make an extensive analysis of the 'Questione Meridionale' (Southern Question), focusing on the antithesis between town and countryside and the deeper coincidence, on a class basis, of interests in urban versus rural areas. In his Prison Notebooks, he argues that:

> In this type of city (old medieval) there exists, among all social groups, an urban ideological unity against the countryside, a unity which even the most modern nuclei in terms of civil function do not escape ... There is hatred and scorn for the 'peasant', an implicit common front against the demands of the countryside - which, if realized, would make impossible the existence of this type of city. Reciprocally, there exists an aversion - which, if 'generic', is not thereby any less tenacious or passionate - of the country for the city, for the whole city and all the groups which make it up. This general relationship is in reality very complex and appears in forms which on the surface seem contradictory; it had a primary importance in the course of the struggles for the Risorgimento, when it was even more absolute and operative than it is today. (A. Gramsci, 1975, p.91)

It is thus not surprising that until the rise of Fascism in Italy there was a persistent regional conflict between the industrial north and the rural south - having its roots in the antithesis between city and countryside as described by Gramsci, an antithesis which in many cases tended to obscure basic class conflicts.

These conflicts became explicitly evident after the end of World War I, when revolutionary ideologies were translated into a series of factory and agricultural land occupations. Proletarian restiveness reached a climax in 1920. The workers in the north demanded a governing share in industrial enterprises (see the discussion of 'factory councils' by Gramsci, 1923) and the peasants in the south demanded fundamental land reform. Both demands, however, were stopped violently by the Fascist 'squadristi' of Benito Mussolini in 1922. The Fascists used uneven regional development between north and south for their own political benefit (Sylos-Labini, 1965). They inaugurated a thoroughgoing and efficient system of repression, especially in northern regions where socialists and communists had greater power. For the south, Mussolini organized a social programme in 1927 for 'Ruralism' aiming to 'ruralize' Italy (ruralizzare), i.e. reconstructing the principles of 'rural life'. Acccording to Sereni (1962), this was more an attempt to move away from northern militant urban centres and to establish Fascist bases of support among southern peasants than an effort to solve the southern question.

Spain

In Spain, the dominance of Catalonia and the Basque region generated similar problems to those of northern Italy. Catalonia, for example, far from being a backward region, had become the only region in Spain to generate a significant industrial surplus product by the end of the 19th century. Spearheaded by an aggressive indigenous bourgeoisie, this industrial development appeared to be on the verge of spreading throughout a nation by the onset of the 20th century.

During the 19th and 20th centuries the south of Spain fulfilled a function comparable to that of southern Italy, not only as an 'internal colony' (Garcia-Ramon, 1978), but also in the sense of the constant reproduction of regional underdevelopment. Above all the Spanish south was used as a catchment area for additional capital, which through a process of transfer of surplus labour was squeezed out of agriculture (Mandel, 1976) to accelerate the process of industrialization in old and new industrial centres in central and northern Spain.

The basic problem during this period remained the north-centre-south conflict, best expressed through the agrarian structure. Southern Spain was predominantly semi-feudal under the control of large land-owners. Northern Spain's agricultural structure was based on small family holdings, and its urban and industrial structure was far more developed compared with that of the south. The centre was under the direct control of the Crown, based on the traditional bureaucratic and administrative role of Madrid. This regional differentiation among north-centre-south directly influenced the social, political and cultural life of the country.

To protect landed interests, the Madrid government placed a civil guard - Guarda Civil - in every village in rural areas. This type of oppression was combined with unemployment and declining wages, both among agricultural and urban workers, resulting from the limited modernization in agriculture and industry at the beginning of the 20th century (Breitbart, 1979).

In addition, the lack of leadership among the peasantry and the weakness of the Spanish state - despite the spread of the Guarda Civil - permitted anarchist ideas to diffuse more easily in Spain compared to other southern European countries. By the beginning of the 20th century anarchist activities were widespread in southern France, in the Mezzogiorno, the Peloponnese and Crete (Moskof, 1974). Although the French, the Italian and the Greek states were able to smash the anarchist movement at its early stage, the Spanish state showed a remarkable weakness and delayed following similar actions. These conditions combined with the low development of the productive forces and the physical characteristics of the country to set the foundations for the strong anarcho-syndicalist movement of Spain and help to explain the anti-authoritarian attitude of Spanish peasants and workers and their belief in regionalism - the right of every region to decide its own destiny - and self-management.[10]

In the countryside (especially in the north among smallholders) where dispersion of population and the strength of village particularism made the establishment of large-scale unions difficult, anarchism and anarcho-socialism became the dominant revolutionary ideologies. In the industrial regions of Barcelona and the Basque country, the syndicalist emphasis on the creation of large-scale revolutionary labour unions proved more fruitful. This dual type of labour self-organization was directly related to regional productive specialization and became important during the years 1914 to 1920 when the cost of living more than doubled. Price increases were spatially more pronounced in southern regions, where the landless peasants found their salaries increased by 20% while the cost of living increased by 85% (Vicens, 1969).

From 1918 to 1920 unprecedented labour agitation - inspired as in Italy by the Bolschevik revolution - shook Spain. A total number of 463 strikes was reported in 1918, 895 in 1919 and 1,060 in 1920 (Garcia-Ramon, 1978). The pattern was the same in agriculture, though the number of strikes was loer because of the lack of coordination: from the total number of strikes mentioned above, 188 were agriculture-based in 1919 and 194 in 1920. As Lorenzo (1969) points out, Spanish agricultural strikes in 1920 exceeded in number (though not in intensity) the rural walkouts in Italy which were to contribute so heavily to Mussolini's rise to power two years later. But neither urban nor rural strikes were very successful, mainly because of the lack of effective nation-wide coordination. Even with the establishment of national organizations like CNT (Confederacion Nacional del Trabajo) in 1910 (especially strong in Catalonia and Aragon) and FNAE (Federacion Nacional de Agricultores de España) in 1913, spontaneous upheaval on a regional level remained the basic form of struggle. This lack of coordination and the internal conflicts within the Republican Army (between anarchists and Stalinist communists) had disastrous effects during the Civil War (1936) which brought Franco to power (Amsden, 1979).

Although anarchists made a series of political and strategic mistakes and finally suffered political and military defeat in 1939, they nevertheless contributed a set of new and promising theories and practices. The programmes of rural collectivization, industrial self-management, decentralization and anti-statism are among the best known (see Mintz, 1970; Bolloten, 1968). Anarchists posed the idea of a 'region' as the most fundamental cell in economic and social life. During the Civil War, the combined economic power of rural collectives in 'regional federations' protected the autonomy of individual villages by aiding them to overcome severe wartime pressures (Breitbart, 1979). An important goal in these regional federations, aside from decentralization, was the establishment of a new relationship between rural and urban proletariats. This was accomplished in part, on the one hand by replacing competitive exchange between 'unequals', with mutual association between urban and rural units, and on the other by controlling the negative effects of indirect geographical transfer of surplus product (in CMP, of surplus value) to the cities (Alaiz, 1937).[11] The process is described by Breitbart (1979):

Recognizing the relationship between economic

> centralization and urban expansion, they also discussed the role cities played in extracting surplus from hinter lands ... The social revolution began to bridge the gulf between peasants and the city workers and to reduce the historical flow of surplus value towards cities ... The most important linkages were established in the direct transfer of food by 'bread committees' in rural communes to urban neighbourhoods ... Urban syndicates provided rural collectives with technical help on water resource projects, labour for the harvest and medical aid ... (Breitbart, p.90)

Greece

In Greece, the regional problem during the end of the 19th and the beginning of the 20th century was primarily one of military and administrative character. Between 1881 and 1918 Greece doubled its territory and population with the annexation of wheat-growing Thessaly in 1881 and later of Epirus, Macedonia, Thrace, the Aegean Islands and Crete, as a result of the Balkan Wars and World War I.[12] This spectacular expansion, in combination with the development of railways and the rapid monetization of the economy, created for the first time a unified and relatively large internal market, a fundamental precondition for the development of commercialized agriculture and indigenous industry (Mouzelis, 1978). The new territories, however, made the problem of agrarian reform a vital one.

After World War I, when annexation of new territories stopped, agriculture developed into a sector of key importance to the regional problem. Turkish landlords sold their *ciftliks* cheaply to rich Greeks of the *diaspora*. These newly established Greek *ciftlik* owners enjoyed, under Roman Law, more privileges and rights over the peasant population than did their Turkish predecessors. Thus the peasants' hereditary right to the cultivation of land, already weakened with the rise of the *ciftlik* system in the 17th century, was now virtually abolished through state actions (Vergopoulos, 1975).

The most serious regional problem regarding population expansion emerged in 1922. After the defeat of the Greek army by Kemal Ataturk's forces, 1.6 million Greek refugees came to Greece from Asia Minor. The desperate need to accommodate this huge mass of people accelerated the land reform started in 1897. By 1936, a total of 425,000 acres had been distributed to 305,000 families (Vergopoulos, 1975). Land reform was appied primarily in central and northern Greece, where ciftliks were still the predominant mode of ownership. At the same time, both the newly arrived qualified labour force which settled in urban centres, and the enormous influx of foreign funds (reaching 1,162.8 million golden francs between 1923 and 1930) brought about the first major breakthrough in the industrial sector. From 1923 to 1939 the value of industrial production doubled and its volume tripled. Industrial activities, however, concentrated in only five major cities: Athens, Piraeus, Patra, Volos and Thessaloniki. The rest of Greece remained an agricultural region. As a result, the gap between urban and rural development increased.

Contrary to other southern European countries and the Balkans, Greece did not experience peasant struggles and a strong peasant political movement at the turn of the century (Mouzelis, 1978; Tsoukalas, 1981). Peasants were brought into the political arena primarily through their dependent integration into the major urban-based political parties, both bourgeois and the Communist Party of Greece (KKE). The development of 'southern peasant socialism' in Italy and Spain can be seen to a large extent as a political and social reaction to the strains and oppression created by the landlords and the increasing penetration of north-western capital, none of which was the case with the Greek peasants at that time.

The failure of the Greek peasantry to organize politically and territorially was to a large extent due to their early involvement in a highly commercialized agricultural production for export (Vergopoulos, 1975). The agrarian reform in 1897 and later in 1909 - twenty years earlier than in Spain and 42 years earlier than in Italy - provided arable land to the majority of Greek peasant families. They thus developed a certain economic autonomy earlier than their fellow peasants in Spain and Italy, but at the same time a greater dependency on dominant capitalist institutions and the state for loans, credit, market regulation, etc.[13]

Like their fellow peasants in southern Italy, Greek peasants were 'outward looking'. On the one hand commercialization of production required contacts and connections with urban consumer centres; on the other smallholdings were hardly able to provide the means of subsistence. Vergopoulos (1975) suggests for Greece that migration rates are correlated with degrees of commercialization and specialization in minifundia: the larger the specialization and commercialization, the greater the chance for economic bankruptcy of small landowner peasants and the greater the potential for migration. Vicens (1969) argues for Italy and Spain that the enormous class distinctions in latifundia created an atmosphere of labour militancy which found expression through revolutionary labour organizations aimed at the destruction of the existing social system. Similar observations are made by Malefakis (1970) for differences in migration patterns between northern and southern Spain.[14] The major point is that in minifundia regions (in northern Spain, northern and parts of southern Italy and Greece) land is fairly evenly distributed; thus no obvious adversary is there against whom protest may focus. Migration remains the predominant response to economic stagnation and social exploitation in these regions.

The period from 1922 to 1935 in Greece was characterized by political instability and growing militancy of mainly urban proletarians. In only a few cases was there joint action of both urban and rural militants, the peasantry being integrated into bourgeois parties (Svoronos, 1972). One of these cases of joint action was the regional strike organized by the KKE in the northern region of Macedonia in June 1936. The tobacco workers in Thessaloniki and Kavala - the two largest cities of the region - joined the walk-out of rural workers protesting against low wages and high costs of living. The confrontation with police and the army ended with more than one hundred dead. These events in the north and others in central Greece and the Peloponnese were exploited by the extreme right. Inspired by similar developments in Italy and Germany, the Fascist dictatorship of Mataxas seized power on 4 August 1936, with a coup d'état. Metaxas, with the support of the Crown, proposed in 1937 a vast 'rural' programme for Greece, '... to put an end to agrarian backwardness of the country', following the Italian example.[15] But the major object of Metaxas' programme was the control of radical workers and peasants in their respective regions. A few years earlier than Franco, he founded an efficient secret police machinery diffused in every village and neighbourhood in the country.

Thus, by the end of the 1930s, Fascist regimes were well established in Spain, Italy and Greece. They publicized both 'nationalism' and 'ruralism' and gave special attention to backward agricultural regions and to decentralization for political reasons. This special 'attention', however, was not an attempt for rural and/or regional development. The Fascists were mainly interested in securing their political base among the small agricultural entrepreneurs. No matter how many objections may be raised against this common treatment of Fascist plans in southern Europe, the fact remains that all these societies passed through that kind of regime and did so under relatively similar historical conditions. Fascist dictatorships appeared when the traditional political formulas of bourgeois parliamentary domination through exclusion of the subordinate classes and parties began to break down (Giner, 1982). They occurred through the pressures of increasingly radical and revolutionary movements, combined with serious problems in the economy. These were beginning to be felt by political orders which were becoming quite unfit for them. It was then that the several varieties of modern despotism which received the name of 'Fascism' came to the rescue, imposing their own order.

Summarizing broadly this brief and selective discussion, three points need to be stressed. The first is concerned with the '*longue durée*' of uneven regional development in southern Europe. It has been the outcome of the gradual development of spatial specialization, of differences in production technologies and modes of labour control and - more importantly I think - of differences in political institutions and modes of political domination. These characteristics differentiated early southern European societies and have been reinforced in later periods. The end result was a concrete and uneven geography of capitalist development, where any new attempt for accumulation was either restricted or enhanced by a continuously changing social and spatial division of labour.

The second point is related with the struggle for benefits. During this long process of differentiation and homogenization, the most elementary struggle (see Wallerstein, 1983) and the most obvious, was that between the small group of great beneficiaries of the system and the large group of its victims. This struggle goes by many names and under many guises. The process of accumulation, however, has been also a struggle *among* the beneficiaries for the distribution of surplus extracted from their victims. The latter led to sectoral and geographical concentration of capital, where the geographical transfer of surplus labour and surplus value played a key role. Of particular importance here is the formation of nation states and the political antagonisms that this process generated.

The third and final point is related to interesting historical differences among southern European societies in terms of the intensity and forms of struggle. In agricultural regions, for example, we have such diverse practices as anarchism and collectivization in Spain, rural syndicalism in Italy and the absence of any significant peasant movement in Greece. This diversity cannot be reduced either to economic factors or to local cultural traditions. What is more important is first, the mode of exclusion/inclusion of rural population into political development in each country and in each region; and second, the relative strength/weakening of local civil society *vis-à-vis* the growing power of the central state. It was at this period that the antithesis between the state and local community started to develop, to become in later periods one major source of regionalism.

NOTES

[1] The physical and climatic unity of the Mediterranean is organized on a long belt straddling the 37th and 38th parallels. The differences in latitude are not great. They are sufficient to explain the contrast between the northern shores and

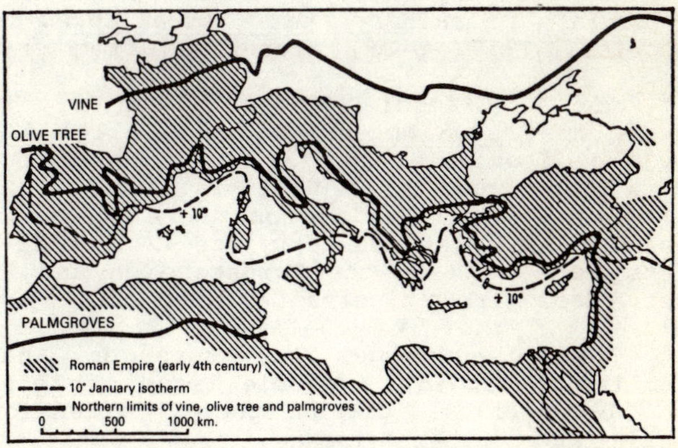

The 'true' Mediterranean, from the olive tree to the great palm groves

the southern, the latter being the warmest. In human terms, the unity of the climate prepared the ground for the establishment of similar rural economies. Source: Braudel (1972)

[2] This system did not lead to the development of capitalist agriculture, i.e. to the creation of a rural proletariat. During the entire period of Ottoman rule, wage remuneration played only a peripheral role in the Greek countryside. See C. Vergopoulos (1975), p.86.

[3] Ambelakia, a region in Thessaly, is the best known example of this type of growth. At the peak of its expansion, it had a capital of 20 million gold French francs and employed between 4,000 to 5,000 people, exporting textiles to Germany, Italy and Spain. See N. Mouzelis (1978), p.10.

[4] This argument is elaborated at length by I. Wallerstein (1974), pp. 236, 266-8.

[5] C. Vergopoulos (1975) emphasizes the fact that when the Greek state decided to distribute national lands among the peasants its main concern was to ensure that the land reform would not result in the concentration of land in private hands.

[6] For Italy, see Indovina (1969).

[7] The special role of the Greek state in shaping the Greek social formation is described by C. Vergopoulos, 'The functioning axis of the Greek social formation was not bourgeois civil society, as a certain liberal theory would imply, but the state. Ever since the middle of the nineteenth century, nothing could be done in Greece without it necessarily passing through the machinery of the state. The state apparatus, as Gramsci would say, was the social machinery *par excellence*.' (C. Vergopoulos, 1975, p.86, translated from original Greek).

[8] See A. Gramsci (1975), 'Notes on Italian History', in *Prison Notebooks* (originally 1929 and 1935). A good overview on the debate about southern Italy is to be found in E. Nocifora (1978).

[9] In 1906 mainland troops were called in to repress the Sardinian peasantry and in 1908 Milanese soldiers were sent to the Taranta area and Sicily against Mezzogirono peasants. See D. Novacco, (1976).

[10] From the large literature on anarcho-syndicalism in Spain, a double issue of *Antipode*, Vols. 10-11 (1979) gives special attention to regional and urban issues: see articles by Breitbart, Garcia-Ramon, Amsden, *et al*.

[11] Myrna Margulies Breitbart (1979, p.93)

describes the process: 'Networks of exchange established by regional federations before 1938 promoted entirely new patterns of circulation in Spain, penetrating the artificial barriers which once isolated rural villages from one another and rural areas in general from the cities. Boundaries between areas were altered, public works projects begun, and transportation systems revamped ... In August 1936, the autonomous government of Catalonia implemented a new set of territorial demarcations based on an earlier perception study by geographers, which attempted to gain a sense of the "real" areas with which people identified (Generalitat de Catalunya, 1933). Similar territorial alterations were made in other anarchist regions to facilitate intercommunal exchange ...'

[13] The agrarian reform consequently caused big land ownership to decline. At the same time it increased the degree of commercialization which had important socio-psychological and political effects on the peasantry. According to C. Moskof (1974) the Peloponnesian peasants, who were involved very early in the market economy, showed much more individualistic, 'modern' characteristics: i.e. less attachment to the soil, greater attraction to the city and willingness to adopt urban styles of life, development of consumer ideology and political attitudes towards the right and the King. See, C. Moskof (1974), p.76.

[14] According to E. Malefakis (1970), one of the major reasons for the Civil War in Spain and the importance of anarchists was the failure of the first Republic Government to apply agrarian reform in the southern regions of the country. Op. cit. note [2].

[15] See newspaper, NEA HELLAS, 12 April 1937

Chapter 7
THE NEW INTERNATIONAL DIVISION OF LABOUR IN SOUTHERN EUROPE

In the decades since World War II world capitalism entered a global reorganization and in the Mediterranean started to reshape itself into a new order, part of a new International Division of Labour (IDL). This new IDL developed in southern Europe around the dichotomy and complementarity of 'Fordism' and 'peripheral Fordism' (Lipietz, 1983b). Fordism was used by Gramsci in pre-war Italy and by Henri de Mas in France to describe the American model or a regime of <u>intensive accumulation</u> characterized by extensive use of machinery, scientific work organization and workers' <u>savoir faire</u>.[1] While France and northern Italy, after World War II and the Marshall Plan, established themselves within 'central Fordism', Spain and Greece (together with Portugal and other 'newly' industrialized countries) fell behind to what Lipietz (1983b) called 'peripheral Fordism'. Peripheral Fordism is characterized by a combination of old style import-substitution with modern export-promoting strategies, industrialization via foreign capital investment, the considerable rise of middle classes, together with state bureaucracy, intensive tourist development, international labour migration and, above all, a specific type of authoritarian rule. In Spain, Italy and Greece, with varying intensity, this new IDL reached dimensions that affected simultaneously the economic, political and cutural/ideological levels.

At the economic level, since the 1970s the impressive growth rates in GDP in the three countries throughout the 1950s and 1960s slowed down (see Table 7.2). A reduction of accumulation and investment took place and resulted in an increase in unemployment and underemployment. These developments did not necessarily affect the whole range of

Table 7.1: Demographic Indicators

	Greece	Italy	Spain
Population (mid-1980s, millions)	9.6	56.9	37.4
Population density (sq.km.)	72.7	189.0	74.1
Population growth (annual average percentage, 1970-80)	0.9	0.6	1.0
Crude birth rate per 1,000 population (1980)	16.0	14.0	15.0

Source: World Bank, 1982

Table 7.2: Gross Domestic Product, 1960-80

	France	Greece	Italy	Portugal	Spain
GDP, 1980 (US$ millions)	651.9	40.4	394.0	24.1	211.1
GDP per caput, 1980 (US$)	12,140	4,210	6,910	2,430	5,650
Growth of GDP, 1960-70 (annual average percentage)	5.5	6.4	5.3	6.2	7.1
Growth of GDP, 1970-80 (annual average percentage)	3.5	4.9*	3.0	4.6	4.0
Percentage GDP from agriculture, 1980	4	14.3	8	13	9
Percentage GDP from industry, 1980	36	33.7	41	46	36
Percentage GDP from services, 1980	60	52	51	41	55

* 1970-79

Sources: OECD, 1982; World Bank, 1982

capitalist enterprises; some privileged sectors remained profitable and continued to expand. Thus, for example, Emmanuel (1977) and Mingione (1978) have shown that the electronics, petro-chemicals and machine tools monopolies were not directly affected by the crisis, while the steel industry, shipbuilding, consumer goods and agriculture were heavily affected. The regional distribution of these effects was very disadvantageous for southern Europe. On the one hand, the sectors that suffered more were mostly located in peripheral regions, and this further deepened their backward position; on the other hand, the monopoly sectors, not directly affected in the first place, were located in core regions within those countries, or outside of them (Secchi _et al_, 1978).

At the political level, economic crisis led to political struggles in which the central question was economic and political democracy, shaped differently in each country. The overthrow of dictatorships in Portugal and Greece by 1974, and the death of Franco in 1975 allowed under the pressure of the masses for a certain degree of democratization of public life in these countries. In Italy during the same period, the consumeristic aspects of the 'economic miracle' came under criticism - all the more remarkable in a country whose poverty was not a distant memory but a contemporary reality (Sasson, 1977). Since 1975 (in Italy since 1970) working class organizations (trade unions, political parties, local committees, etc.) powerfully entered into the political arena in all the countries, following a growing radicalization and a shift of the masses to the left. The new leftist movement soon began to connect the conditions of exploitation in the factory with exploitation in a wide range of other levels and encompassed urban and regional questions, health and education problems, environmental issues, and the women's movement. It also came to recognize the need for a level of government wider than the municipalty and yet closer to territorial life than the central government (Zangheri, 1975; Bretonnière, 1976; Ingrao, 1976).

Finally, at the ideological/cultural level, the impact of economic and political crises increased the need for legitimation of the administrative system and more generally of state action. The high degree of economic and political involvement of the state in Italy, Spain and Greece (followed by a remarkable inefficiency in many aspects) made legitimation an indispensable factor <u>vis-à-vis</u> the radicalization of the masses.[2] With the collapse of dictatorships in Spain and Greece and the inefficiency of the Christian-Democrats in Italy, the whole ideological hegemony of the bourgeoisie in southern Europe faced a direct threat.

The masses were no longer willing to restrict themselves to purely economic demands, since these could not be met within the framework of the existing Spanish, Italian and Greek economies. The working class, radical peasants, students and sections of the middle class began to challenge the very process of capitalist development itself and examine alternatives for southern Europe (Poulantzas, 1976).

The result, after 1974, was an authentic crisis of hegemony and few areas of collective life escaped the attention of the radical masses. In this respect, uneven regional development became a major arena in the struggle for alternative development in southern Europe for three main reasons:

a. It challenged a key problem of national development common in all three countries: spatial over-concentration of activities, people and wealth in a few core regions and a widening gap with the rest of the regions. Table 7.3 illustrates this process of concentration in terms of GDP.
b. It emphasized the need for regional decentralization and self-management at the local level as basic components of the left wing struggle against Fascism and bureaucratic centralism.
c. It provided a viable and appealing alternative to the historical tendencies for regional autonomy in Spain and Italy and for cultural regional differentiation in Greece.

Thus, by the mid 1970s, the process of capitalist centralization - economic, political and spatial - interrelated with local historical traditions of autonomy, gave birth to two expressions of regionalism in southern Europe: one a regionalism that moved away from the separatist movements of the beginning of the century and their current counterparts and focused on economic, political, cultural and environmental issues (Tarrago, 1976); and the other, a regionalism founded on nationalism having strong historical and cultural roots in certain areas, demanding complete

Table 7.3: Spatial Concentration of Production (Percentage of Total GDP)

	Spain		Italy		Greece	
	1955	1975	1951	1976	1951	1975
5 top provinces having highest GDP per capita in 1955 or 1951	34.5	39.5	26.2	31.8	35.8	42.4
10 top provinces in 1955 or 1951 (including the previous 5)	46.0	50.4	37.0	43.2	48.3	56.6
5 bottom provinces in 1955 or 1951	5.1	4.1	10.2	9.3	4.8	3.1
10 bottom provinces in 1955 or 1951 (including the previous 5)	9.9	8.4	13.6	13.3	6.3	5.2
Remaining provinces	44.1	41.2	49.4	43.5	45.4	38.2
	100.0	100.0	100.0	100.0	100.0	100.0

Source: OECD, 1975; EUROSTAT, 1976; Greek National Accounts, 1978 to my calculations

autonomy, leading to separatism.

7.1 CHANGES IN THE SOCIAL STRUCTURE AND THE SPATIAL DIVISION OF LABOUR

While the dichotomy of central and peripheral Fordism has some merit for an overall descripton of the position of each southern European country within the IDL, it is nevertheless inadequate in describing internal regional and social variations. If we wish to investigate the changing social and spatial relations such a description seems to be too abstract and functionalistic (see also Cooke, 1983). A basic inadequacy of this type of approach is that it takes as the basic unit of analysis the nation state. Social classes are viewed solely as national and the focus is on the male work-force and the absolute and relative rates of mobility in and out of such classes.

In southern Europe since the 1950s, the process of rapid industrialization, the development of infrastructure and intensified state intervention, resulted in a certain degree of national economic integration across regions. This very process, however, generated (at least until the 1980s) side effects of severe social and spatial differentiation due to changes in local production and exchange relations and in patterns of local political domination. I now want to suggest that, because of these changes in contemporary capitalist relations in southern Europe, regional and local class structures are in fact of increased importance. 'Regional' and 'local' are used here (as in Chapter 5) to refer to labour markets defined in terms of travel-to-work areas (Paci, 1978) and the articulation of various forms of production particular to a place (Garofoli, 1983). One could argue against this, that in a period of interregional and international hypermobility of labour and capital in southern Europe, the appropriate unit of class analysis is that of the nation state or even the EEC as a whole. I do not believe, however, that this is the case, and I will try to show why (see also Mottura and Pugliese, 1975; Garofoli, 1983; Tsoukalas, 1984).

The following five points are relevant, serving to highlight some of the important changes in the last 30 years which are significant for regional and local class structure:

1. The crisis that took the form of a gradual

reduction of the role of the landed agrarian bourgeoisie, resulted in southern Europe in its overall liquidation as a governing class at the national level. Nevertheless its importance remains high at the local level in certain regions, in combination with urban or tourist speculators (e.g. southern Italy, Sicily, central Greece, Crete, Spain's Mediterranean coast, parts of the Meseta). Land speculation became the easiest and most profitable path for indigenous economic success in southern Europe. Land fragmentation and small properties permitted a large number of medium and large capital holders to become developers, contractors and tourist entrepreneurs; they formed a powerful class fraction (especially in Spain, southern Italy and Greece) with important political influence (see Graziani, 1975).

2. The liquidation of the landed bourgeoisie was not followed by a general pattern of development of the industrial bourgeoisies on the national level. Long before this liquidation took place, certain regions, due to specific historical conditions of accumulation, developed important endogenous industrial bourgeois classes (e.g. northern Italy, central Greece, northern Spain). These regional industrial élites, together with the commercial bourgeoisie in old dominant trade cities, became national classes, but they nevertheless preserved their regional identity. An interesting exception here is the Greek commercial bourgeoisie of the eastern Mediterranean and especially its shipowner fraction - what Poulantzas (1974a) calls a comprador bourgeoisie - which became dominant in the post-war period in Greece without a regional base inside Greece.
3. Following the development of state apparatuses a 'new' class or class fraction emerged: the state petty-bourgeoisie (Poulantzas, 1974a). Pugliese (1979), writing for Italy argues that already under Fascism the appointment of so many small bureaucrats to the state machinery gave them the illusion of massive participation in state power, while the real power remained in upper state officials and the regional industrial bourgeoisies. The same was true for Francoist Spain and Greece during the 1967-74 dictatorship.[3] Increase of low and medium

ranking state personnel also occurred after 1950 in Spain, Greece and southern Italy related to high rates of urbanization and 'industrialization without employment'.

The state sector and other tertiary activities absorbed a large part of the population without, however, increasing efficiency and quality of services (Tsoukalas, 1981). With the growth of public economic intervention under authoritarian rule, these new social strata acquired special local strength through control of the local public spending process and the local authorities. A new set of social relations and dependencies was established at the regional level around the control of employment policy, administration of public works, management of agricultural and industrial credit, welfare and pension funds, etc. All this spending was distributed mainly via patronage and clientelism (not to speak of loyalty to the dictators), providing additional sources for regional social class differentiation.

4. The apparent weakness of the national industrial working class (as percentage of the total labour force) also has limited explanatory power, when it is not broken down to regions and localities. Traditional militant working class centres, such as Torino, Milano, Barcelona, Bilbao, Aragon, Piraeus, Hermoupolis, Volos, are in sharp contrast with newly industrialized regions, where a new kind of working class with less solidaristic attitudes is developing. On the regional level, conditions change dramatically, according to the size and internal make up of the proletariat and above all its political importance. As many labour market studies have shown, of major significance is the rapid development of new forms of production based on 'domestication' of the labour process (working by piece at home, part-time jobs in the informal economy etc.). In Italy, this productive decentralization strategy is mainly promoted by big and medium size industrial firms, like FIAT. These changes signify a dynamic process of restructuring in relation to the antagonism between accumulation and workers' resistance. In Thraki and in the western part of the region of central Greece, for example, productive decentralization is applied by newly

located small firms that took advantage of regional incentives (Hadjimichalis and Vaiou, 1985). These firms (mainly in textiles) after one or two years in operation closed production and distributed their equipment to family units for in-home by-piece production (see, newspaper AVGI, 21 March 1984). In contrast with northern Italy, Thraki is a less unionized region, and firms have distributed their production for economic reasons, including the lack of disputes with labour.

5. Finally, a considerable reduction and an important internal alternation of the peasant masses has taken place throughout southern Europe. I will have the opportunity to discuss this process in detail in the next two chapters. For the moment it is worthwhile noting that the central figures are no longer the poor peasants and the farm workers in latifundia, but a much more elaborated hierarchy of rural social figures (Mottura, 1980). The current picture again reveals important regional variations of social class organizations. In the lowlands and coastal regions, where industrial and tourist development was concentrated, a combination of modernized farming and part-time jobs outside agriculture provide an adequate income. In the stagnant interior, by contrast, proletarianization took the form of impoverishment of the peasantry and a higher dependency on the state for subsidies and pensions. The peasants from these regions provided the majority of emigrants; millions of them became industrial workers inside their country (northern Italy, Athens and Thessaloniki regions, northern and the coastal regions of Spain, etc.) or outside in Switzerland, France, West Germany, Belgium, and so on.

Summarizing briefly the preceding discussion, it seems that the tendency towards economic homogenization across regions has been followed by an even greater tendency towards social differentiation and the formation of new class distinctions across regions. Thus, regional social classes and labour markets are increasingly important elements for an adequate understanding of regional unevenness in southern Europe.

Three major phases in their post-war

development can be identified. First, the capitalist restructuring that took place after World War II (in Francoist Spain, in liberal Italy and in right wing anti-communist Greece) brought about a modification of the structure of the state and the configuration of social relations. The decades of authoritarian rule introduced a massive state intervention and a new choice for rapid economic development through the so-called 'open gate' policy, to foreign capital investment. This choice was confronted with an allocation of resources in favour of monopolistic industry which had to be paid by all segments of the population. Secondly, the development pattern in the three countries during the crisis period between the 1960s and 1970s was based on the use of ever increasing amounts of labour force, in efficient and spatially concentrated big plants with low wages, and resulted in widening regional disequilibria (Secchi, 1972; Arcangeli, 1982; Teixidor and Hebbert, 1982; Kafkalas, 1981). This period was marked by a reinforced hegemony of the leading sectors of Spanish, Italian and Greek capitalism. It brought about a process of increasing homogeneity/ differentiation, through progressive 'modernization'.[4] And thirdly, since 1979 the new development pattern seems to combine widening and deepening of regional inequalities, making the traditional north-south distinction highly problematic. Inequalities are developing now among and within regions depending on a system of localized social relations and an increasingly fragmented regional labour market. Within the context of a prolonged international economic crisis, a certain degree of productive decentralization is taking place. With important variations from country to country, these changes coincide with localized crisis in 'old' industrial centres and changes in technology, management and capital labour relations (Garofoli, 1983, Hudson and Lewis, 1984).

7.2 ITALY

In Italy, the general economic crisis that followed World War II required not only an immediate search for the means of economic reconstruction but also basic decisions regarding the role that the state could play. Under these conditions the role of planning, both sectoral and national, came out clearly (Palombara, 1966). From a regional point of view the major planning actions were land reform, decentralization of industry and formation of the 'Development Fund for the South', known as Cassa per il Mezzogiorno (Holland, 1971).

Land reform was one of the major demands of Socialists and Communists in the immediate post-war period. The Christian Democrats (DC), however, chose not to pass legislation until 1956 that could bring about land reform throughout the country but instead to limit expropriation to particular regions, especially in the south, where latifundia existed (King, 1973). Even this limited implementation of land reform could never have been initiated without the violent peasant struggles that characterize the last years of World War II and the immediate post-war period. These struggles, marked by the southern peasants' demand for land, forced the hegemonic element in the Italian class structure - the northern bourgeoisie - to sacrifice the interests of the landowning southern bourgeoisie, in the same way as the Greek state did during the last part of the 19th century. Up until 1955, reform legislation affected only 29.3% of the total national agricultural land in Italy. About 8% of that portion went to those who were already small landowners; 40% was assigned to tenant farmers and share-croppers; the remaining 52% went to agricultural day-labourers (Saraceno, 1973).

Industrial decentralization - advocated by the Socialists - was another very important issue. The damages of war were much greater in the Mezzogiorno than in the north, where industries remained almost untouched (Donolo, 1972). In addition, IRI (Institute for Industrial Reconstruction), a large state holding company created during Fascism, was actually involved, as Pastore (1966) mentions, in a large-scale programme to finance northern industrial development at the relative expense of the south. This process was exaggerated by post-war reconstruction handled by the DC in a manner that clearly favoured large industrial concentrations in the north.

The Cassa per il Mezzogiorno, established in 1950, was the first massive effort in Europe for regional development through a specialized agency. The main goals of the Cassa at the time of its creation were to promote agricultural development in the south through land reform projects of various kinds, loans and technical advising. Furthermore, it aimed at improving infrastructure and tourism and, to a lesser degree, industry (Saraceno, 1973). In fact, 77% of initial investments were intended for agriculture, 20% for infrastructure and 3% for tourism; in 1966 the allocation was as follows: 55% for agriculture, 30% for infrastructure, 3% for tourism and 12% for industry. The Cassa's programmes, together with those of the independent research Institute of Industrial Development of the South (SVIMEZ), have been heavily criticized by most planners because of their reliance on infrastructural policy only and their limited grants for industrialization (Chenery, 1968; Garofoli, 1974; Holland, 1971). In effect the Cassa was to be the means of implementing a regional development programme. It would create the infrastructure assumed sufficient for the location of industry and the development of agriculture in the south to occur through the 'normal' working of the market mechanism. In practice, the government failed to support the infrastructural programme of the Cassa when it became evident that indigenous industry in the south was not responding to the available incentives and that agriculture and industry located in the north similarly were failing to respond to both the available financial incentives and the newly provided industrial infrastructure (Saraceno, 1973).

A consciousness of the impeding limits to and the 'dualistic' character of growth emerged only at the beginning of the 1960s at the peak of the 'economic miracle' (Arcangeli, 1982). Sectoral and regional disequilibria were interpreted only as 'limits to growth' and not at all as functional elements in this specific growth model (Secchi et al., 1972). Until the end of the 1950s major institutional reforms were initially proposed but then abandoned, due to DC opposition. Moreover, no real integration of the special policies introduced by the Cassa per il Mezzogiorno and national planning policies was achieved (Graziani, 1975).

The general development pattern in the 1960s was characterized by concentration in space of large capital-intensive plants, which were designed to reduce the high technological gap existing in Italy after the war (Garofoli, 1983). This strategy had important effects in the regional labour market which in turn - through the class struggle in 1962-63 - changed capital's strategy towards a more diffused industrialization (Paci, 1979). Until 1963, wages were sufficiently low to enable Italian regions to compete internationally without making new investments. The labour market ought to have stayed 'loose'. Moreover, industrial production was concentrated in a few areas in the north, making the depressed areas in the south even poorer.

The situation changed radically around 1963 when important bargaining strikes developed, organizing class struggle on a massive scale. For the first time since the war, the increase in wages was higher than labour productivity gains, and this was a real victory for the Italian working class. Since that time Italian capitalist development has taken different forms. It is no longer an extensive development encompassing bigger and bigger proportions of industrial production and employment but it is becoming a 'deep', 'intensive' development. The regional labour market becomes a 'buyer's market' once more, and the labour demand becomes 'selective' according to certain characteristics (sex, age, health, etc.). These trends have been analysed by many scholars including Paci (1979), Garofoli (1974) and others.

Secchi (1977) in his study of regional inequalities in Italy (1950 and 1970) identifies two phases: during the first (1950-1958) what he calls 'extensive' (i.e. an extensive utilization of the labour forces) regional differences in terms of net per capita income grew considerably, while the aggregate rate of increase of employment was positive (see Table 7.4); during the second period (1965-1970), what he calls the 'intensive' phase, regional differences were still modestly increasing (in absolute figures they were lower compared to 1958 but still higher compared to 1952), while there was a decrease in employment.[5] Secchi emphasizes the fact that, except during the decline in 1965, inequalities in terms of gross product per permanent worker increased significantly over the entire period due to the continued industrial concentration in northern regions.

Thus, according to Secchi, regional inequalities in post-war Italy were functional for the development of Italian capitalism and in his words:

> ... the existence and growth of regional inequalities made the Italian economic system more flexible in terms of labour supply than it would have been in a better balanced regional situation, given an equal rate of employment in the various sectors and regions of the economy. (Secchi, 1977, p.36)

Table 7.4: Regional Inequalities in Italy

	1952	1958	1963	1965	1969
1)	.2960	.3372	.3370	.3210	.3233
2)	.0951	.1286	.1188	.1148	.1072
3)	.1382	.1493	.1538	.1480	.1615

1) = net per capita income at market prices
2) = total number of employed in all sectors
3) = gross product per permanent worker at market prices

Source: Secchi, 1977. See [5] for formula used to calculate values in table.

The developed pattern after 1970 became more complex. Dualistic interpretation (north versus south) no longer held and new ones have arisen, emphasizing the multiple and not dual structure of the Italian economy (Bagnasco, 1977). As Arcangeli (1982) noticed, the traditional division into big regions (north, centre, south) is no longer important because the local differences (within each region) are assuming at least the same importance as the external ones.

This evolution from north-south dualism towards a more diffused pattern of uneven regional development should have required much more effective regional planning, but even in present-day Italy regional planning means little in practice, because of the limited power of regions over their own expenditures. This was a strange development in a country which (together with Spain) has a progressive constitution promoting regional self-government.

The 1947 Constitution of the Italian Republic recognizes and promotes decentralization and regional autonomy. However, it was only after twenty-two years, in 1972, that a government act validated it. Two years earlier, in 1970, the Italians elected regional councils for the first time, in an atmosphere of growing radicalization. The demand for regional self-government enacted those articles of the Constitution dealing with regions and acquired characteristics of a demand for reform of the entire state machinery. As Sassoon

(1977) points out:

> Concurrently, the peasantry's demands for increasing local autonomy were no longer solely based on a distaste for a central authority present in the community in the form of policeman, judge or prefect ... The new regionalism of the peasantry was increasingly oriented towards problems of economic policy. (D. Sassoon, 1980, p.16)

In these circumstances the PCI conducted a powerful campaign for the regional elections of 1970. It proposed the term 'Open Region', following Gramsci's notion of 'new revolutionary historical bloc'.[6] The Open Region constituted a new form of territorial self-administration based on a new social majority: a coalition of all democratic forces (except the neo-Fascists [MSI]), for co-responsibility in local affairs and regional development initiated from below. The opening of a region's affairs to all people living in it was PCI's contribution to decentralization and an antidote to the centralized and closed system proposed by the Christian Democrats and parts of the Socialist party.

The application of this programme in the regions and municipalities - the 'red areas' - where the PCI and PSI took the majority of votes was an important experience in Italy. At that time (1975) the country was particularly badly hit by the economic crisis, and this generated a general debate on social policies. The crisis afflicted many cities and regions in the north and south. The innovative administration of the red areas - particularly in Emilia-Romagna - then acquired great importance. As a result, in the general and regional elections of 1975 the PCI increased its power by 5.59% and the PSI by 1.56% of the votes.

But the electoral victories of the left at the regional and municipal level would never have materialized or would have remained meaningless without an active, mass social movement. An indication of this mass mobilization around territorial issues (urban, rural, regional) in Italy is, on the one hand the direct participation of people in a large number of councils, committees, action-groups, etc., dealing with territorial issues; and on the other their direct involvement in street-action (demonstrations, strikes, occupations, etc.).

This participation started as early as 1948 with the '<u>Consulte Popolari</u>' (The Popular Advice Centres) organized by the Resistance, which quickly spread throughout Italy, in neighbourhoods, villages and the countryside. The <u>Consulte Popolari</u> were independent from trade unions and were responsible for developing a struggle at the place of living. They had been working since 1965 with factory committees, which organized actions concerning improvements in working and living conditions. The individual <u>Consulte Populari</u> became associated at the national level in 1970 and, from a number of 9,813 centres in 1972, they increased to 32,021 in 1977.

These developments, especially since 1975, changed the social, political and territorial structure of Italy. The regional reforms and the mass mobilization introduced the concept of 'region' as the basic unit of direct democracy. As Carassi (1975) pointed out:

> ... the Italian regions have become the main field for struggle against Fascism, and our experiences from them can guide the way towards their socialist transformation. (A. Carassi, 1975, p.86, my translation)

Popular democratic struggles at the regional scale, however, did not succeed in establishing self-management and never got regional control of finance and spending. They nevertheless succeeded, together with the working class struggles at the end of the 1960s, in challenging the old dualistic development pattern and in breaking up the alliance of the old leading forces of Italian capitalism.

Thus, in the 1970s the traditional poles of industrial concentration within the 'industrial triangle' lost momentum, and new areas of economic dynamism in the centre and north east expanded steadily. This process was followed by de-industrialization and de-urbanization in major northern cities such as Milan, Turin and Genoa. At the same time peripheral demographic growth along the two coasts started to replace the old depopulation pattern (Dematteis, 1983). Finally, the most typical Italian spatial pattern in the 1970s was the process of 'diffused industrialization' (see Table 7.5). It involves a productive decentralization based on small firms (from 9.3 employees per plant in 1971, to 7.4 in 1981) and formation and development of local productive systems (Garofoli, 1983). Productive decentralization in Italy - as well as in other southern European countries - took advantage of certain local economic and social characteristics such as fragmented and discontinuous regional labour markets, local resources, local entrepreneurial skills and locally introduced technologies.

7.3 SPAIN

In Spain, the period from the end of the Civil War to the 1950s was characterized by economic stagnation, autarchy and political oppression (Tamames, 1978). Official sources indicate a decline of per capita real income from 8,500 pesetas in 1935 to 5,400 in 1945, at 1953 money values, which of course involved a much greater fall in real wages (OECD, 1974). Between 1945 and 1950 the cost of living increased by 60% while wages remained blocked. It was only after 1950 that there was a gradual recovery of real wages, which probably reached their 1935 levels only towards the beginning of the 1960s (Carr and Fusi, 1981).

After early 1957, economic recovery and progressive 'modernization' under the Opus Dei movement towards a bureaucratic authoritarian state took place.[7] New policies were introduced at that time, the most important of which were '*desarrollismo*' - the pursuit of prosperity - combining massive US aid, relative wage increases in 1956 and 1958, and reform in foreign trade controls (Rodriguez, 1972). All these were included in the 'Stabilization Plan' put into action in 1959. Another important element in the plan was the new incentives for foreign capital investment that contributed to rapid economic expansion after 1962 (Tamames, 1978).

In 1961, an IBRD study mission visited Spain but made only marginal references to regional development policies in its 1962 report.[8] As Lasuen (1969) and Richardson (1975) argue, emphasis on regional planning came from various regions that traditionally enjoyed autonomy, Catalonia and the Basque region being extreme examples, and Navarra, Galicia, the Levante, the Balearic Islands and the Canary Islands more moderate ones.

Traditional nationalisms were reinforced by high economic and social differentials across regions generated to a large part by increasing spatial concentration of both population and industry in Madrid, Barcelona, the Basque country and Valencia. For example, two provinces, Barcelona and Madrid, accounted for one-fifth of total industry in the country in 1950 and more than two-fifths in 1970. If the Basque provinces and Valencia are added, the concentration rises to over 56% by 1970 (Teixidor and Hebbert, 1982).

The First Development Plan (1964-67) did give some attention to regional problems despite the fact that the World Bank's guideline was to subordinate regional aspects to the overall goal of faster aggregate growth. The major goal adopted by the Plan was a balanced development of all regions. This involved the creation of poles of industrial promotion and development in low income regions, establishment of conditions favourable for internal migration, and agricultural modernization.[9] In this respect two industrial promotion poles (Burgos and Huelva) and five industrial development poles (La Coruña, Seville, Valladolid, Vigo and Zaragoza) were designated.

The growth pole strategy was to be the dominant component of Spain's regional policy up to 1975.[10] Since then, however, questions have arisen as to whether this policy has succeeded in 'spreading' development in the poles' hinterland. Richardson (1975) for example argues that the poles have shown considerable 'backwash effects' and points out that:

> ... even where the pole and the province appeared to thrive side by side, the pole has grown at the expense of the province. (p.131)

The period from the end of the Civil War to the death of Franco was marked by repression of the various Spanish regional cultures. This was the outcome of the centralist and authoritarian character of the Francoist state, which enjoyed the support of the monopoly corporate sector and the economic oligarchy, including those from the repressed regions. Those fractions of the bourgeoisie were in favour of a unified and expanded national market under their hegemony and a rapid integration of the Spanish economy into the world capitalist market.

The Fascists were especially oppressive against the northern regions of the Basque country and Catalonia.[11] In the Basque country they even prohibited the use of the local language, music and Basque names, while in Barcelona, among other things, an attempt was made to destroy the regional culture through a well defined educational system

(Perez-Agote *et al.*, 1978; Benedito and Fuentes, 1978). Under these conditions a growing radicalization of the masses took place, and in the Basque country during 1959 the militant organization ETA (*Euskadi Ta Askatasuna* - Basque Country and Liberty) was founded and began an armed struggle against the Franco regime by 1967.[12] ETA was involved in a great variety of actions and activities, ranging from cultural festivals to bombing and from political demonstrations to pacifist negotiations. It was widely supported by the local working class, the peasantry and by the non-monopoly fraction of the local bourgeoisie, which was interested in a wider class alliance on a cultural-territorial basis against the Madrid government. Perez-Agote (1978) argues that the so-called 'Basque movement' and, to a lesser degree, the 'Catalan movement' for regional autonomy and preservation of local culture, must be considered as 'nationalist' movements (i.e. mobilization of distinct cultural and territorial groups with a separatist tendency) and not as 'regional' social movements as the Spanish state describes them. From this point of view, the Basque and the Catalan movements have more characteristics in common with the Corsican nationalist movement (United Front for the Liberation of Corsica, UFLC) than with, for example, the Andalusian mobilization (1967-77), the Calabrian Popular Association for Development (1908 to the present) or the various Cretan Local Cooperatives for Development, established since 1974 (to be discussed in detail in Chapter 11). The latter would be considered as regional movements in that they question the domination of those regions by the central state and their backwardness (not only economic) without imposing the issue of separatism and independence.

The boom of the 1960s transformed Spain from a predominantly agricultural country into the world's ninth largest industrial power. However, the highly uneven pattern of regional development between industrial and rural areas and the high percentage of the population active in agriculture still emphasizes the importance of this sector in regional performance (Jesus, 1970). The development of Spain during the boom years was among the most imbalanced in Europe. The partial diffusion of growth in the late 1960s and early 1970s was due to acute bottlenecks of infrastructure in the leading areas, and to the shift of mobile industry, facilitated by regional incentives.

The death of Franco in 1975, the introduction of a new Constitution in 1978, the return to democratic municipal elections in 1979, and the current devolution of regional autonomy have created a very different framework for regional development. The 'transition', however, as in the case of Greece, did not represent a full break with the previous regime. It left intact key institutions of the dictatorship within the civil service, the armed forces, the mass media and the economic system (Tamames, 1978). Nevertheless, it was a period of general political euphoria for democracy and most political parties were brought together by the worsening economic situation and fears of an army coup d'état.

Thus, after the re-establishment of democracy, the question of regional autonomy emerged again. Limited autonomy was granted to Catalonia and the Basque country by 1977. The Catalan parliament, the Generalitat, was re-established for the first time since 1938. In the Basque country, a regional council composed of members of parliament and regional representatives was also formed. Demands for regional autonomy were expressed in demonstrations throughout November and December of 1978 in Andalusia, Navarra, Seville and the Canary Islands, often turning into running battles with the police.[13]

New planning attempts include a revision of the 1956 Land and Planning Act in 1975 to its new 'Planes Directores Territoriales de Coordination'. Unlike earlier regional plans, which were merely consultative, the new ones are statutory documents with legal powers, and they form a hierarchy of spatial plans, with the 'Plan National de Ordenacion' at the top (Teixidor and Hebbert, 1982).

As was the case in Italy, under the pressure of the masses a major regional reform was proposed through the new Constitution of 1978. The Constitution clearly outlined the movement towards a federal structure, but the country's regions were to have only limited autonomy (Rodriguez, 1972). Thus, the principal demands of the two major leftist parties - the PSOE (Socialist Workers Party of Spain) and the PCE (Communist Party of Spain) - in the regional and municipal elections of 1979 centered around an increase of regional autonomy and the allocation of more direct administrative responsibilities to territorial units. The left gained important victories in all major urban areas and in the more developed regions, while the Centrist Party was able to take majorities only in some rural municipalities and low-density regions (Le Monde, 4 April 1979).

A principal effort of the new leftist administrations was the extension of their power. They had to struggle against the persistence of an authoritarian and centralist ideology which has dominated Spain since 1939. Thus decentralization and regionalism became arenas of intense fighting for the stabilization of democracy and for an active cooperation of all democratic forces with an anti-Fascist and anti-imperialist orientation (Borja, 1976).

In this effort, as M. Tarrago (1976) argues, the regional movement in Spain was supported by the rich tradition of self-management and regional autonomy movements of the Civil War. In addition, it had incorporated the urban social movements which, from 1969 to 1975 and within the limits imposed by Franco's dictatorship, gained important victories in Spanish cities. These urban movements have been able to establish a number of citizen organizations, such as the '*Juntas Municipales de Distrito*', which were transformed into militant nuclei in the last years of Franco and the first years of democracy. The basic problem in Spain remains, however, to establish a broader alliance between the various regional leftist self-organized councils (especially between urban and rural areas), a difficulty which goes back to the Civil War and which will most likely characterize the Spanish regional movement for the years ahead.

In summary, the continuous economic growth until the mid 1970s was highly regionally uneven, concentrated in Madrid, Catalonia and the Basque country and some of the larger cities, such as Zaragossa and Valencia. Foreign investment played a key role especially in chemicals, metal goods, vehicles and food processing sectors. Both emigration and tourism contributed to growth with remittances and tourist receipts covering about 80% of the trade deficit. In contrast, the agricultural sector remained less developed until 1969-70, when significant advances in commercialization and mechanization took place.

7.4 GREECE

When the occupying forces withdrew from Greece after the end of World War II, leftist resistance forces controlled the greatest part of Greek territory. They were none the less driven into a Civil War for which they were not prepared and suffered complete military defeat.[14]

As was the case in the Spanish Civil War, the leftist forces in Greece from 1943 to 1949 were involved in reshaping life in the rural and mountainous regions. A large programme of land collectivization was carried out successfully in Thessaly and Macedonia; agricultural and small-scale industrial production was organized on a regional basis; educational, health and cultural facilities were established in a number of villages (Eudes, 1970). During this short period, regions and communities in Greece enjoyed extensive autonomy, self-reliance and decentralization of power (Svoronos, 1972). This was the reason why the left found widespread support from the rural populations in backward regions, despite the relative conservatism of the peasantry one decade before, when Metaxas came to power.

The end of the Civil War found Greece almost ruined: 9,000 villages, 23% of all buildings and 65% of industrial establishments were destroyed. In 1950 the government, under the guidance of the US Military Mission, organized the National Board of Reconstruction which generated a Long-run Redevelopment Programme.[15] This programme was concerned only with the rebuilding of settlements and their infrastructure, without other major development objectives. However, World War II and the Civil War had strongly influenced the regional distribution of the population. The Civil War in particular, which was mainly carried out in the mountains, resulted in the economic destruction of a significant number of rural communities producing migration into urban centres.[16]

The main economic effect in the 1950s was the increased use of foreign aid (mainly from the US) and greater privileges to the private sector. The foreign advisors pushed the government to change the pre-war development policy (which was based on industrialization via import substitution) and to give priority to agriculture. For US planning experts at that time, Greece had 'comparative advantages' only in tobacco, olives, commercial agriculture and tourism.[17] In addition, a heavy emphasis was given to infrastructural investment, mainly in highways and dams for electricity.

These new policies had important effects on spatial organization, such as the expansion of agricultural land through mechanization and the shift of priorities from railway to highway investments. Without major industrial investment and new job opportunities in rural areas, however, the mechanization in agriculture, poor rural living conditions and the new road system resulted in accelerated urbanization.

The most striking economic characteristic of the post-war period was the spectacular growth and concentration of finance capital and its tight control over the whole of the economy. A very complex process of mergers and takeovers resulted in the emergence and consolidation of a duopolistic situation in which the National Bank of Greece and the Commercial Bank of Greece - now both owned by the state - control about 96.3% of the assets of all Greek commercial banks.

Despite the impressive rate of growth in GNP - an average 6% in the 1950s and 5.6% in the 1960s - the Greek economy did not manage to overcome a major difficulty: its weak, industrial structure. Until the mid 1960s Greek capital, ever searching for higher profits and lower risks, preferred to orient itself, mostly on borrowed money, towards the non-manufacturing sectors, especially commerce, construction and tourism. Given these structural weaknesses, and the state's long-term commitment to a free-enterprise economy, there was no other solution for Greece than to resort to the help of foreign capital. The share of foreign capital in GNP increased from 2.15% in 1962, to 8.15% in 1972 and 11.6% in 1978. Both industrial production and productivity of labour tripled their respective performance from 1958 to 1968 (Germidis and Delivanis, 1975).

The impressive sectoral and spatial concentration of industrial and finance capital did not eliminate the large number of small and medium size (employing 1-9 people) industrial and commercial units, which were widespread along the so-called 'S' development corridor (Patras-Corinth-Athens/Piraeus - Lamia - Larisa/Volos - Thessaloniki-Kavala) and which for the most part are of family or extended family character (see Table 7.5) Indeed, one of the most striking characteristics of Greek industry and the corresponding regional labour market is the persistence of these small units side by side with large firms. While in Italy the growth of these small units was more rapid after 1963 as an outcome of the restructuring process, in Greece it has been an endemic phenomenon since the end of the Civil War. Until the 1970s these small units were mainly concentrated in the more traditional sectors (footwear, clothing, leather, wood products, etc.) and operated with low productivity; since then, however, their productivity and specialization has increased rapidly to include modern sectors, such as aluminium products, plastics, synthetic textiles and fur processing plants, mainly in northern Greece.

Table 7.5: Small Firms in Southern Europe and Industrial Employment

Firms employing	Number of employees 1-9 %	10-49 %	50-99 %	Over 100 %	Average Plant size	% of industrial employees in total salaried employment (1981)
Greece (1979)	91.3	6.5	1.6	.6	5.2	28.5
Italy (1981)	84.7	12.7	1.4	1.1	7.4	31.9
Portugal (1971)	78.9	15.9	2.6	2.7	15.6	33.9
Spain (1970)	76.6	18.5	2.5	2.4	15.0	32.9

Sources: World Bank, 1982; ETVA, 1980; OECD Labour Statistics, 1983

In April 1967, following an unstable political situation, a military coup d'état established a seven year dictatorial rule in Greece.[18] Contrary to what happened in Spain and Italy, and with Metaxas in 1936, the April dictatorship lacked any popular base when it seized power in 1967 and was unable to mobilize the masses to give a Fascist character to its rule. By following the logic of the economic model it had inherited, the junta gave its unlimited support to big foreign and domestic capital. It made sure through repression that the ensuing growing inequalities would be accepted unconditionally, without any protest or strikes to frighten capital away (Poulantzas, 1975). Thus, the rate of growth soon surpassed pre-dictatorial levels and sustained an impressive acceleration of industrialization and tourist development.

Throughout the decades of the 1960s and the 1970s the general standard of living undoubtedly rose. The $500 per capita income in 1950 rose to $2,070 in 1975. But this was achieved at the cost of amplifying inequalities. For instance, Germidis and Delivanis (1975) estimated that in 1973 40% of the lowest income groups received only 9.5% of national income (after deduction of taxes), whereas 17% in the top income brackets received 58%. Another study by Karagiorgas (1980) on national per capita income found the Gini coefficient rising from 0.39 in 1950 to 0.41 in 1960, 0.45 in 1971 and 0.46 in 1976.

The social concentration of wealth was followed by spatial concentration as well. A study by the Ministry of Coordination and Planning showed that in 1974 56% of total industrial units (plants employing 10 people or more) were in the Athens region, employed 58% of the total national labour in secondary activities and produced 49% of total value added in the industrial sector. In addition, in 1977 the Athens region accounted for 30% of total population, 56% of the GNP, 51% of total private fixed capital investment in all sectors and 31% of total state investments. In the same period Thessaloniki region, in Macedonia - Thessaloniki being the second largest city of Greece - accounted for 16% of GNP, 13% of the total private investment and 12% of the total state investment (NNS, 1977).[19]

In a way, the military dictatorship that seized power in 1967 was the first victim of the world recession in 1972-73. The latter brought the expansion of the Greek economy of the 1960s to an end and deprived the dictatorship of its momentum. Politically, the junta's foundations were too shallow for it to survive this downturn. In 1974, after the massacre in the Athens Polytechnic and the Cyprus crisis, the junta, under the pressure of the rising social revolt, yielded its power to a transitional civilian government. The transitional government of 1974 legalized the Communist Party, which was split in two since the late 1960s: pro-Soviet and Eurocommunist, but it left intact most authoritarian institutions developed during the seven year dictatorial rule. The new Constitution approved by the Parliament in 1976 did not give the same attention to regional problems as those in Italy and Spain. While the problem of geographical concentration and regional inequality seems more severe in Greece, the government did not pay direct attention to it. During the last three years of dictatorship, a significant number of social mobilizations occurred, questioning urban and regional inequalities. Issues such as housing and transportation, prices and distribution of agricultural products, environmental protection, etc., increasingly became the focus of social conflict. It was not only increasing social and spatial stratification that generated these conflicts, but also the obvious inability of the military regime first to handle the popular uprising and secondly to provide viable solutions.[20]

With the re-establishment of democracy these social movements have acquired momentum, following the growing radicalization of the population. As was the case in other southern European countries, in the municipal elections of 1975 and 1979 the left achieved the majority in all important cities, including Athens, Thessaloniki and Patra.[21] But existing legislation gives little room to leftist forces for intervention at the municipal level and excludes them totally from regional administration. In this respect, regional and municipal reform for self-government have become important areas for political struggle (see newspaper AVGI, 4 April 1978).

The three major leftist parties, however, have not yet managed to organize an alliance between all democratic forces at a territorial level in order to struggle effectively for common territorial development. The self-organized 'Provincial Popular Committees' soon became a focus for political conflict when the various parties competed to control them, each for their own political benefit, thus destroying the popular initiative and confining the struggle to short-run political disputes.

To conclude this section on the new international division of labour (IDL) and the countries of southern Europe during the post-World War II period until 1979, three points could be made. First, at the national level the most notable political development was the collapse of dictatorships in Spain and Greece and the growing inefficiency of DC in Italy to handle the structural crisis of Italian economy. Political responses during the same period in the three countries vary considerably, but nevertheless they have one thing in common: a rise of popular struggles followed by a rise of the power of leftist parties. In addition, labour migration, capital movement and economic restructuring have their impact in increasingly differentiating the nation and regions of southern Europe, which now occupy qualitatively different positions within the new IDL.

Secondly, at the regional level an important development was the distinctive locational pattern associated with particular forms of production. Its effect has been to differentiate and at the same time to homogenize regional economies, regional social classes and regional labour markets. This pattern has on occasion provoked regional social movements within each country, which combine economic, cultural and political factors. An important dimension of these regionalist reactions was their relation with the general leftward swing in southern European politics, especially at the local level.

Thirdly, the new forms of regional specialization in agriculture and industry, new forms of tourist development oriented to the mass international market and international migration and capital movements cannot be understood without reference to the European Community. However, it is beyond the scope of this study to make a full reference to general EEC policies; the latter require a different research focus. Nevertheless, the increasing tendency to regional specialization in the three countries has reflected a domestic choice to seek economic links with the EEC, with a special emphasis in their agricultural sector. This sectoral emphasis was combined with attempts for regional development. Thus, in the next three chapters I will present an empirical example of uneven regional development and geographical transfer of value based on the analysis of the agricultural sector in Italy, Spain and Greece.

NOTES

[1] Gramsci's well known work 'Americanism and Fordism' is concerned with the impact of American productive methods on Europe. He also introduced 'Taylorism' as the ultimate rationalization of production and - for the first time - he identified the internal differentiation of the working class along production lines - see <u>Prison Notebooks</u>, 'Americanism and Fordism', pp.275-318.

[2] It has been estimated that in 1975 state intervention in the economy through the banking sector in southern Europe was as follows: Greece 90%, Italy 75%, France 65%, Portugal 63% and Spain

43% - see OECD (1979).

[3] Of special importance was the role of the army and police personnel. In Spain and Greece these class fractions acquired - and still preserve - special power, enjoying social and economic privileges (e.g. social status, higher wages, special housing, shopping centres and vacation facilities, etc.).

[4] A continuing large outflow of southern labour, however, attracted by the rapid industrialization in the north and the abandonment of large areas in the south, resulted in a waste of public investment in agriculture and infrastructure and was among the factors causing a major change in development policies by about 1962. See also, Holland (1971).

[5] The various indicators used by Secchi have been calculated with the formula:

$$V_y = \frac{[\Sigma i (yi - \bar{y})^2 pu/p]^{1/2}}{\bar{y}} \quad (i = 1, 2, 3, 4)$$

In each case a four-region territorial disaggregation has been used: north-west, north-east, central and southern Italy. For a similar methodology and calculations, see also Garofoli (1974, 1976).

[6] Gramsci employed the term 'revolutionary historical bloc' to refer to a historically-congealed synthesis of popular movements. By 'bloc' Gramsci meant considerably more than simple alliance and coalition. What he envisaged was an amalgamation of forces, always shifting and changing, that emerges at a specific historical conjuncture. In the Prison Notebooks, he discussed 'national blocs', urban blocs', 'southern rural blocs', in which different social groups come together as movements through a shared ideology. See also, Mouffe (1979).

[7] Opus Dei was the technocratic modernized élite within the Francoist tradition. Its major function was to help Franco's generals cope with difficult economic problems and to negotiate with US technocrats and foreign capital.

[8] The World Bank's aspatial argument was that the general solution of the problem of regional imbalances is to be found only in a policy designed to stimulate the overall growth of the national economy. According to the Bank, a high rate of growth, reinforced by measures for promoting the free mobility of labour and capital, constitutes the most efficacious means for raising standards of

living in rural areas. See, World Bank (1966).

[9] Since 1969, when the first Development Plan came into effect, Spain's regional policy has adopted the 'Industrial Poles' as its direct instrument in action.

[10] Under the French influence of F. Perroux and J. Boudeville, the Spanish experience with growth pole planning is the most extensive in Europe. The poles were selected without consideration of regional economic needs and were viewed primarily as a spatial instrument for promoting national industrialization. This suggests, as Richardson (1975) argues, that the poles were primarily oriented towards the national and international market, and secondarily towards the regional one. In this respect, the expectations for 'spread' effects towards the regional economy proved to be very optimistic.

[11] The Spanish Basque provinces include Alava, Guipuzcoa, Navarra and Vizcaya, while the French Basque provinces include Laburdi, Zaberoa, Benavarre. The Basque country and Catalonia were centres of the Democratic Army during the Civil War. Franco considered the Basque provinces during and after the Civil War as provinces 'to be punished', starting with the total destruction of their historical capital, Guernica.

[12] ETA was organized initially in three sections (political, cultural and military) and was in collaboration with the Basque Nationalist Party, in exile at that time. In 1973, a Trotskyist section of ETA split and founded the <u>Liga Communista Revolucionaria</u> (LCR) and in 1976 ETA faced its major split in two fractions: the first, the ETA-politico-military, accepted the status of autonomy decided after negotiations with the Madrid government; the second, ETA-military, did not accept any autonomy given from Madrid and demanded complete freedom.

[13] See <u>El Pais</u>, 13 December 1978.

[14] By the time of German withdrawal, the EAM - the Leftist Resistance Patriotic Front - and the ELAS - its military/partisan section - controlled over 70% of the Greek territory. The intervention of the British, however, and a series of crucial mistakes made by EAM and the Stalinist KKE leadership, ended in an unequal Civil War. See D. Eudes (1970).

[15] See, the role of National Board of Reconstruction, described by McNeill (1978).

[16] During the Civil War, rural populations in the mountains were forced by the governmental army to migrate to cities, to prevent ELAS recruitment of additional democratic forces. See Eudes (1970), op. cit[14].

[17] Op. cit., note[15] .

[18] In the debate on the rise and fall of the dictatorship, the CIA is often represented as an omniscient and omnipotent force regulating and controlling everything. Now there is no doubt that the CIA, both before and after 1967, had strong links with officers and the Greek state. But this alone cannot explain the coup d'état in 1967, nor the change in 1979. According to N. Poulantzas (1975), behind the rise and fall of the dictatorship was the conflict between two fractions of the bourgeoisie: the 'indigenous bourgeoisie' related to European monopolies and the 'comprador bourgeoisie', dependent on American capital. See, N. Poulantzas (1975).

[19] Hadjimichalis and Vaiou-Hadjimichalis (1979), using the coefficient of variation for a number of indices, such as GNP per capita, income per capita, population density, industrial employment, housing conditions, education and health, argue that uneven regional development in Greece was intensified in the period from 1937 to 1974. Additional support to these findings comes from two EEC reports, 'On the Enlargement of the Community' (1974 and 1977). The Regional Commission of the EEC points out that:

> ... among the major difficulties that Greece will face in relation to the Community Policies we can distinguish three as the most significant: (a) regional inequalities, (b) agrarian and industrial structure, (c) the differences in the legislative structure. (EEC, 1974/1977, p.12)

[20] For a documentation on the urban and regional social movements in Greece, see Bulletin of the Association of Greek Architects and Planners (1974).

[21] The three major leftist parties in Greece are: PASOK, The Panelehik Socialist Movement, KKE, the Communist Party of Greece (pro-Soviet) and KKE Esoterikou, the Communist Party of Greece of the Interior (Euro-communist).

Chapter 8

AGRICULTURAL TRANSFORMATION AND THE PEASANT QUESTION

In the post-war period, the contribution of agriculture to GDP and as a source of employment has fallen for all three countries, but the importance of agriculture remains significantly greater than that for the EEC as a whole (see Tables 8.1 and 8.2).

Table 8.1: The Importance of Agriculture in Southern Europe

	Contribution of Agriculture to GDP %		Agricultural Employment as a % of total Employment	
	1960	1980	1960	1980
Italy	19.0	8.0	32.8	12.8
Spain	20.0	9.0	42.3	17.4
Greece	24.9	14.3	57.0	28.5

Sources: OECD, 1969; Yearbook of Labour Statistics, 1981

Thus, for the period under study, performance in agriculture directly influenced regional accumulation, regional incomes and living standards. In addition, agriculture in these three countries had a number of characteristics - with important variations from region to region - which made it highly vulnerable to market fluctuation, resulting in sharp regional differences in production and

Table 8.2: Basic Statistics for Italian, Spanish and Greek Agriculture (1975)

Indicators	Italy	Spain	Greece	EEC
1. Total area (Million ha.)	30.1	50.5	13.2	152.3
2. Cultivated area (Million ha.)	14.9	20.7	3.9	94.6
3. Percentage of cultivated area irrigated	24%	11.4%	18.4%	41.3%
4. Cultivated area per capita (ha.)	.26	.62	.44	.37
5. Cultivated area per farm-worker (ha.)	4.60	5.90	2.80	7.30
6. Percentage of farming population in total labour force	17.1%	24.5%	31.2%	9.5%
7. GAP/GDP in percentage	9.8%	12.7%	15.6%	5.7%
8. Annual growth GDP	5%	5.8%	5.6%	5.3%
9. Annual growth GAP	1.5%	3.5%	2.1%	1.8%
10. Value added per farm-worker as percentage of the national average	56.5%	46.8%	49.6%	68.9%
11. Tractors per 100 ha. cultivated land	4.8	1.8	2.7	5.1
12. Fertilizers per ha. cultivated land (in kg.)	102	67	88	90.5
13. Gross fixed asset formation in agriculture as percentage of total	7.3% (1963) (prices)	7.2% (1971) (prices)	11.4% (1958) (prices)	4.1% (1971) (prices)
14. Agricultural exports/total exports	8.5%	23.6%	35.2%	12.2%
15. Agricultural imports/total imports	20.1%	15.7%	9.8%	21.7%

Sources: OECD, 1976. GAP = Gross Agricultural Product. GDP = Gross Domestic Product

consumption patterns. These characteristics derive from the specific mode of integration/ marginalization of agriculture in the social formations of the three countries and the economic and political effects that such an integration/ marginalization implies.

To analyse the agricultural transformation in southern Europe we must pay attention to both capitalist market and state intervention. Such analysis also implies taking sides in the active debate on the appropriate characterization of peasants: whether they form a separate social class, how they behave economically and politically, what are their production changes and whether GTV legitimately applies to southern European agriculture. In this chapter, before presenting a review and an interpretation of this debate, I will first look to some data describing the performance of the agricultural sector in Italy, Spain and Greece.

8.1 PATTERNS OF AGRICULTURAL DEVELOPMENT

Notwithstanding the deep differences in social, cultural and economic conditions, the Mediterranean countries of southern Europe bear in common a number of fundamental problems and patterns of economic change. Since World War II, many of these communalities have been concentrated in the agricultural sector and in the transformation of social and spatial structures in rural areas around the Mediterranean basin. The common problems and characteristics are discussed first in this section, while differences will be presented for each country separately.

Table 8.2 summarizes some basic information for the period under study on Italian, Spanish and Greek agriculture compared with EEC average statistics. From the table we can see that Greece has a greater dependence on its agricultural sector and a greater self-sufficiency in agro-products (higher percentage of farming population, higher GAP/GNP ratio, higher percentage in agricultural exports and lower in imports). Italy is less dependent on agriculture and at the same time shows higher figures in terms of productivity (value added per farm-worker) and capital/land ratio (irrigated land and use of tractors and fertilizers). Spain seems to be in the middle, being less dependent on its agriculture compared with Greece, but having a significantly lower capital/land ratio and productivity compared with Italy. Finally, the EEC as a whole is less dependent on agriculture than the three Mediterranean countries, showing a sharp difference in terms of capital/land ratio and productivity.

The contrast between EEC figures and those for the three countries provides a first approximation of the qualitative differences between north-west and southern European agriculture. This is further developed in Table 8.3, with information from Holland, France, West Germany, Ireland and Portugal. The most capital-intensive agricultural sector is to be found in Holland and West Germany. Holland, which is a major competitor of southern European agriculture for fruit and vegetables (cultivated intensively in greenhouses), shows the highest capital/land ratio, but also a relatively high land/labour ratio due to its family-based high productivity agricultural structure (Eurostat, 1979). West Germany and France also have more capital-intensive agricultural structures but corresponding per capita productivity is lower compared with that of Italy, Spain and Greece.

Central governments and local authorities in the three countries have played a very important role in production and investment, while they also provide their farmers with heavy protection against foreign competitors. Two basic methods exist by which intervention to guarantee prices may be achieved. First, the supply of agricultural products may be curtailed, thus forcing up their prices. Secondly, production and trade may be allowed to determine output and prices, with the consequence that the latter will be 'too low', so that farmers have to be compensated for low prices by being given government subsidies. It means that consumers receive food at low prices and taxpayers pay for the policy. The Greek government guarantees minimum prices for the main agricultural products. Price stabilization is coupled in many cases with direct and indirect income aids which are intended to keep consumption prices relatively low. In addition, almost all Greek farmers are totally exempt from direct taxation (up to 1.5 million drachmas). In Spain, minimum guaranteed prices by the state cover about 60% of the total agricultural production. In contrast to Greece, fruit and vegetables are not included.

Within Italy, changes in agriculture have been heavily influenced by the CAP (Common Agricultural Policy) and despite the latter's controversies,

Table 8.3: Agriculture in Europe-Selected Countries (1975)

	Cultivated land in million ha.	Percentage of cultivated land irrigated	Number of tractors per Km^2 cropland	Fertilizers tons used per Km^2 cropland	Farm-workers per Km^2 farmland	Per capita productivity in agriculture
Holland	2.2	69%	20	73	13	141
W.Germany	10.3	39%	15.6	52	5	112
France	18.8	29%	7.8	24	7	100
Ireland	2.8	16%	7.5	35	5	118
Portugal	4.2	14%	1.4	6	16	103
Italy	14.9	24%	4.9	12	15	114
Spain	20.7	11.4%	1.8	8	9	127
Greece	3.9	18.4%	3	12	17	153

Sources: OECD, 1974; Eurostat, 1979

Italian agriculture enjoys a strong system of intervention and external protection (Tsoukalis, 1981). In the EEC (before Greece became a full member), 99 per cent of its total production of olive oil, 91 per cent of peaches and 72 per cent of tomatoes came from Italy, which all favoured price support from the CAP. There exists, however, a bias of the CAP price-support mechanism in favour of temperate products which widen regional differences in agricultural incomes between northern and southern Italy. Nevertheless, the support system works to deal with problems arising in the market, including also 'withdrawal prices".[1]

Crop Specialization

In terms of the structure of agricultural production and specialization, all three countries show marked similarities in the composition of their production, with a definite emphasis on the production of Mediterranean crops (Papayannakis, 1978; Tsoukalis, 1981). The EEC has defined Mediterranean agriculture as one in which durum wheat, rice, fruit and vegetables (including citrus fruit), flowers, wine, tobacco, olive oil and sheep meat represent 40 per cent or more of the total agricultural production. The only products which are grown exclusively in the Mediterranean regions of the Community are durum wheat, rice, olive oil and some fruits.

More important perhaps than the portrayal of a static picture of agriculture in these countries is an analysis of trends in production over the last decade. Table 8.4 shows the indices of agricultural production of the principal products in the three countries for the period 1969-1981, using 1969 as the base year.

For Greece, the biggest increases in production have been registered with respect to most fruit and vegetables, especially apricots, peaches and tomatoes. Large increases in the production of cereals, wheat, barley, potatoes, poultry and eggs have led to the country becoming almost self-sufficient in these products, though considerable shortages of beef, milk products and fodder grains still exist (Agricultural Bank of Greece, 1980). The regions which benefited most from these improvements were Crete, the Peloponnese and west and central Macedonia. In Spain, the most remarkable increase in production over the last decade have been with respect to all citrus fruit (except oranges) and secondary for peaches, tobacco, wine, apricots and tomatoes. Again regional variations were important,

Table 8.4: Indices of Agricultural Production 1969-1981 (1969 = 100)

	Greece	Italy	Spain
Cereals	154	116	97.6
Wheat	147	91	71
Rice	88	108	114
Barley	121	305	120
Potatoes	136	79	120
Olives	149	102	52
Tomatoes	171	125	123
Onions	111	112	52
Peaches	243	143	189
Lemons and limes	137	92	384
Oranges	146	114	80
Tangerines/clementines/ mandarines	116	120	218
Apricots	281	115	128
Wine	124	112	134
Tobacco	139	156	158
Sheep ('000 head)	104	115	80
Beef and veal (slaughtered)	81	102	111
Mutton and lamb (slaughtered)	110	120	101

Source: FAO, Production Yearbook 1981

favouring Spain's Mediterranean coast, parts of the Meseta, the Basque country and Catalonia. As in both other countries, in Italy barley and the fruit and vegetable sectors have experienced sizeable production increases, often as a consequence of increase in acreage. For other Mediterranean crops, such as wine, tomatoes, olives and tobacco, increases in production levels have been often recorded as a result of new planting and the application of intensive methods. Furthermore, production trends must be seen in the light of the operation of the CAP, which provides varying degrees of support for groups of products. Italian agriculture, in comparison with that of the other two countries, has experienced larger changes over the last three decades, especially in terms of a move away from mixed farming towards a trend of increasing regional specialization in the production of particular products by intensive methods (Pugliese, 1982).

These trends could be better understood when we take into account the national and regional distribution of Agriculture Surfaces Utilized (ASU) and regional crop specialization. A detailed and most useful study made at the 'Institut Agronomique Méditerranéen de Montpellier' by Judez *et al*. (1979, provides a set of comparative data for this purpose. Table 8.5 shows that in all three countries cereals and Agricultural Surface Under Trees (AUT) account for more than 50 per cent of their total agricultural surface. Furthermore, in Greece, AUT alone accounts for 57 per cent of its ASU. In the three countries the lowest percentages are shown to be citrus. Greece also shows lower surface percentages in fresh and dry fruits, and is behind Spain and Italy in terms of surface proportion in such 'traditional' Mediterranean crops as olives and wine.

Maps 8.1, 8.2 and 8.3 provide a first approximation of the regional distribution of crop specialization using data from Judez *et al*. study.[2] The regionalization uses existing regional statistical boundaries. Spain shows (as expected, due to its size and important physical variations) the most geographically diverse crop specialization, while Greece seems to be more homogeneous. In southern Spain, the 'provincias' of the Mediterranean coast from Barcelona to Almeria show a clear orientation towards vegetables, flowers and fresh fruits. The same is true for the Mezzogiorno, Sicily, the Peloponnese and Crete. In the north of the three countries cereals, pasture-ground and AUT seem to predominate. In central Italy (six 'regioni'), the historical pattern of a mixed production is kept with a high diversity.

Agricultural Population

In 1981, 6,106,000 people in the three countries continued to depend on farming for their livelihood. There has, however, been a substantial decline in the proportion of agricultural employment over the last 30 years. With industrial development both within and, more importantly, outside southern Europe, the move away from agriculture turned into a massive rural exodus. There is, though, an important distinction to be made between rural exodus and agricultural exodus, with the latter not necessarily implying movement out of the area. The geographical aspects of rural exodus reflect regional differences in terms of the combination of factors discussed previously and are shown in Maps 8.4 to 8.9 and

Table 8.5: Agricultural Surface Utilized (ASU)

	Spain (1972)		Italy (1975)		Greece (1975)	
	Surface (1,000) (Ha.)	% in total area	Surface (1,000) (Ha.)	% in total area	Surface (1,000) (Ha.)	% in total area
Cereals	7,196	25.7	5,124	23.58	1,548	16.6
Fallow land	5,036	17.7	n.d.	n.d.	501	5.4
Industrial crops	1,136	4.0	364	1.67	300	3.3
Horticulture & flowers	475	1.7	559	2.57	156	1.7
Tuber (potatoes, carrots, etc.)	389	1.4	180	0.83	57	0.62
Grassland	1,215	4.3	4,219	19.42	385	3.7
AUT - Area Under Trees	7,226	25.4	5,244	24.13	5,313	57
Vegetables and grain	675	2.4	317	1.46	105	1.12
Vineyards	1,740	6.1	1,508	8.8	203	2.2
Olives	2,207	7.8	2,171	10	573	6.2
Citrus	228	0.8	205	0.8	46	0.5
Dry fruits	531	1.9	659	3	56	0.6
Fresh fruits	415	1.5	778	3.6	71	0.7
Total	28,469	100	21,720	100	9,315	100

Sources: Judez _et al._ (1979) Institut Agronomique Méditerranéen de Montpellier

Tables 8.6 to 8.8.

The tables are fairly self-explanatory, but it is worth highlighting some of the main points. The figures show a considerable decline in terms of agricultural population in almost all the regions and provinces of the three countries. The only exceptions are the provinces of Guipuzcoa and Lugo in Spain, and Lasithi and Rethymnon in Greece (both in the island of Crete), where there was a small increase. As expected in capitalist economies that

Map 8.1: ITALY - Distribution and Regional Grouping of Dominant Agricultural Surface Utilized (1975)

Source: Judez et al. (1979), Institut Agronomique Méditerranéen de Montpellier

Map 8.2: SPAIN - Distribution and Regional Grouping of Dominant Agricultural Surface Utilized (1972)

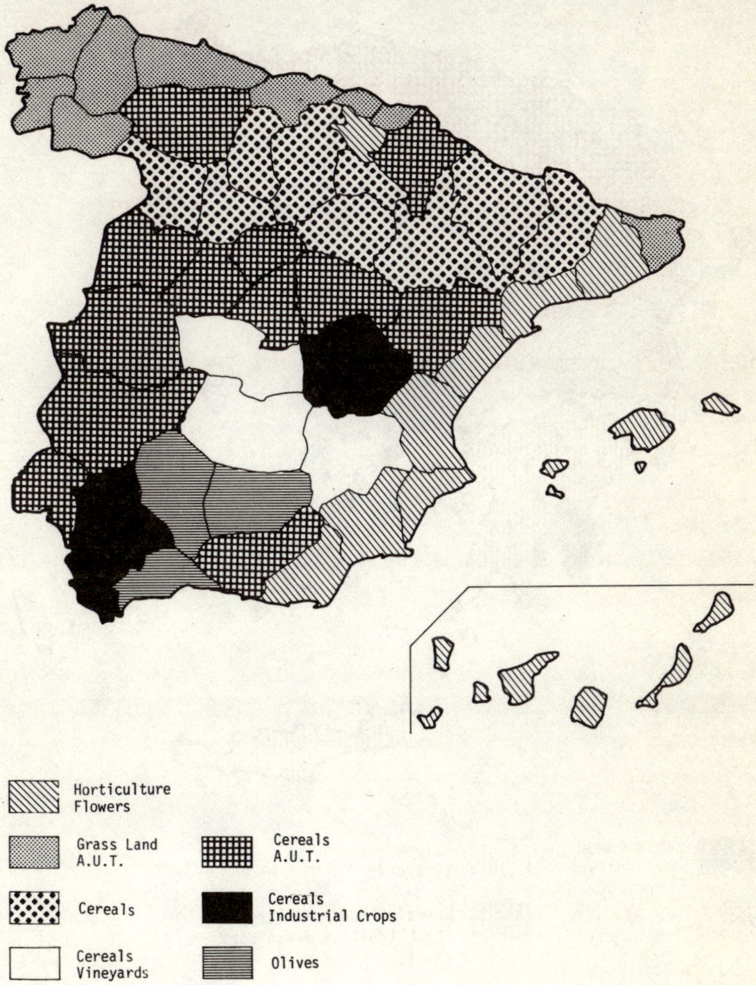

Source: Judez et al. (1979), Institut Agronomique Méditerranéen de Montpellier

Map 8.3: GREECE - Distribution and Regional Grouping of dominant Agricultural Surface Utilized (1975)

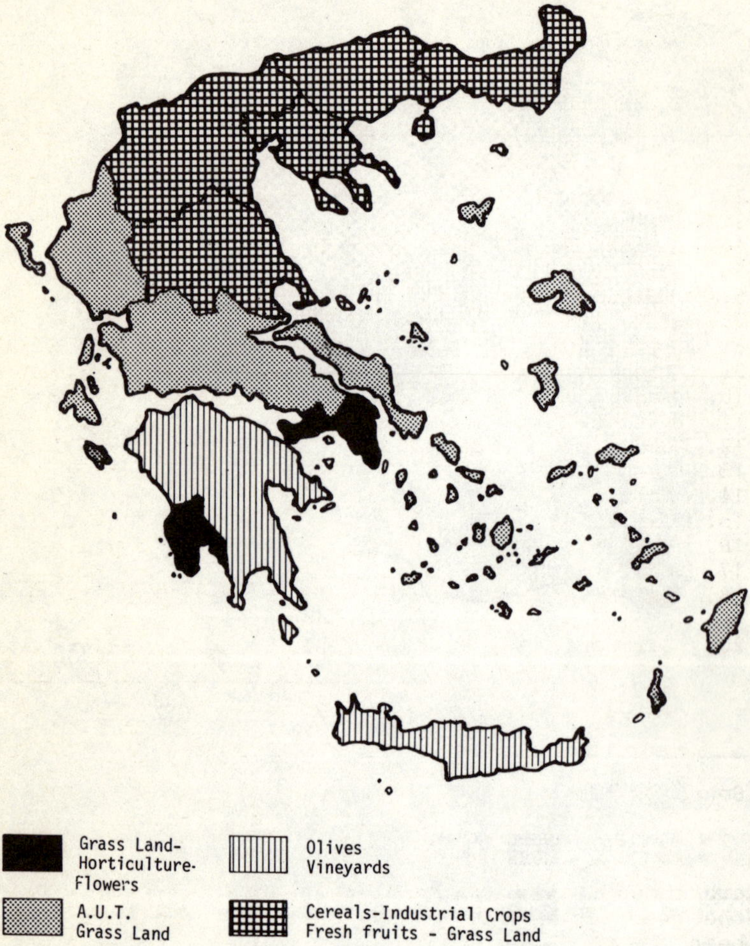

Source: Judez *et al*. (1979), Institut Agronomique Méditerranéen de Montpellier

Table 8.6: Changes in Rural Population in Italy

		Percentage of agricultural population in total active population		Difference 1951-1977
		1951	1977	%
1.	Piemonte	25.7	11.2	-14.5
2.	V. d'Aosta	21.5	13.0	- 8.5
3.	Lombardia	21.2	4.7	-16.5
4.	Liguria	23.8	16.3	-12.5
5.	Trentiono	22.1	12.8	- 9.3
6.	Veneto	15.6	8.4	- 7.2
7.	Friuli	17.7	7.9	- 6.8
8.	Emil. Romagna	28.9	16.8	-12.1
9.	Toscana	25.7	10.1	-15.6
10.	Umbria	23.5	14.9	- 8.6
11.	Marche	36.6	19.9	-16.7
12.	Lazio	26.7	10.6	-16.1
13.	Abruzzi	33.0	23.9	- 9.1
14.	Molize	53.0	43.2	- 9.8
15.	Campania	35.2	24.0	-11.2
16.	Puglia	51.9	33.2	-18.7
17.	Basilicata	56.2	39.7	-16.5
18.	Calabria	43.2	30.7	-12.5
19.	Sicilia	38.6	26.1	-12.6
20.	Sardegna	35.8	19.3	-16.5
	ITALY	33.1	15.7	-17.4

Sources: INEA, 1955, 1965, 1975; OECD, 1975; Annuario di Statistiche Provinciali, 1953, 1963, 1975; Eurostat, 1970, 1979

are industrializing rapidly, the spatial pattern of the general decline in agricultural population has been very uneven. This implies substantial differences between regions and provinces in their employment and productive structure.

Greek prefectures show higher percentages of agricultural population decline but only two Greek prefectures out of 44 - the Athens region and Thessaloniki - have agriculture employment below 10

Map 8.4: The Italian Regions

Sources: INEA, 1955, 1965, 1975

Map 8.5: Italy: Decrease of Agricultural Population, 1951-1977

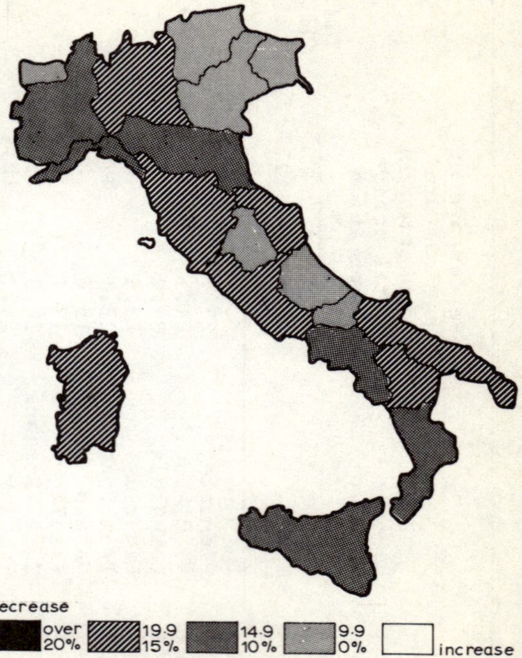

Table 8.7: Changes in Rural Population in Spain

	Percentage of agricultural population in total active population		Difference 1955-1975
	1955	1975	%
1. Alava	14.5	9.2	- 4.8
2. Albacete	24.2	14.6	- 9.6
3. Alicante	19.8	6.1	-13.7
4. Almeria	21.8	15.1	- 6.7
5. Avila	29.0	17.9	-11.1
6. Badajoz	25.0	13.1	-11.9
7. Baleares	17.1	8.7	- 8.4
8. Barcelona	3.0	2.0	- 1.0
9. Burgos	21.5	13.2	- 8.3
10. Caceres	27.6	13.5	-14.1
11. Cadiz	15.5	6.2	- 9.0
12. Castellon	28.5	10.6	-18.0
13. Ciudad Real	24.1	12.6	-11.5
14. Cordoba	25.3	12.8	-12.5
15. Coruna (la)	25.2	18.3	- 6.9
16. Cuenca	26.9	16.9	-10.0
17. Gerona	16.1	8.2	- 7.9
18. Granada	23.1	13.3	- 9.8
19. Guadalajara	26.0	7.4	-18.6
20. Guipuzcoa	6.6	7.1	+ .5
21. Huelva	18.5	9.5	- 9.0
22. Huesca	23.3	15.4	- 7.9
23. Jaen	27.8	14.4	-13.4
24. Leon	25.2	18.5	- 6.7
25. Lerida	24.4	12.3	-12.1
26. Logrono	23.2	10.6	-12.6
27. Lugo	29.6	36.0	+ 6.4
28. Madrid	3.3	.9	- 2.4
29. Malaga	21.1	9.4	-11.7
30. Murcia	19.9	8.4	-11.5
31. Navarra	21.5	7.9	-13.6
32. Orense	36.5	29.2	- 7.3
33. Oviedo	17.1	12.0	- 5.1
34. Palencia	21.3	8.8	-12.5
35. Palmas (las)	22.7	8.7	-14.0
36. Pontevedra	27.7	21.2	- 6.5
37. Salamanca	21.1	9.7	-11.4
38. S.C. Tenerife	23.4	11.6	-11.8

Table 8.7: Changes in Rural Population in Spain (continued)

	Percentage of agricultural population in total active population 1955	1975	Difference 1955-1975 %
39. Santander	18.8	12.8	- 6.0
40. Segovia	19.5	11.7	- 7.8
41. Sevilla	16.8	7.3	- 9.5
42. Soria	27.7	11.6	-12.1
43. Tarragona	24.3	10.1	-14.2
44. Teruel	28.5	13.8	-14.7
45. Toledo	26.6	11.6	-15.0
46. Valencia	20.2	5.4	-14.8
47. Valladolid	18.3	5.3	-13.0
48. Vizcaya	7.3	2.4	- 4.9
49. Zamora	28.2	18.5	- 9.7
50. Zaragoza	17.9	6.7	-11.2
SPAIN	28.9	24.5	- 4.4

Sources: Annuario de Estadistica Agraria, 1965, 1977; Banco de Bilbao, 1975; OECD, 1970, 1975

per cent in 1971. From a geographical perspective (see Map 8.9) faster declining provinces in Greece have a proximity to large urban areas (Beotia, Corinth and Argolis near Athens; Arkadia and Ahaia near Patra; Halkidiki near Thessaloniki). These provinces also practise relatively highly capital-intensive farming.

In Spain the declining pattern seems to be more uniform, and the majority of provinces lost from 7 per cent to 13 per cent between 1955 and 1975. In contrast to Greece, twenty Spanish provinces (out of 50) have agricultural employment below 10 per cent in 1975 and all but two under 25 per cent show significantly lower figures in terms of agricultural population compared with those for Greece and the Italian south.

Map 8.6: The Spanish Provinces

Map 8.7: Spain: Decrease of Agricultural Population, 1955-1975

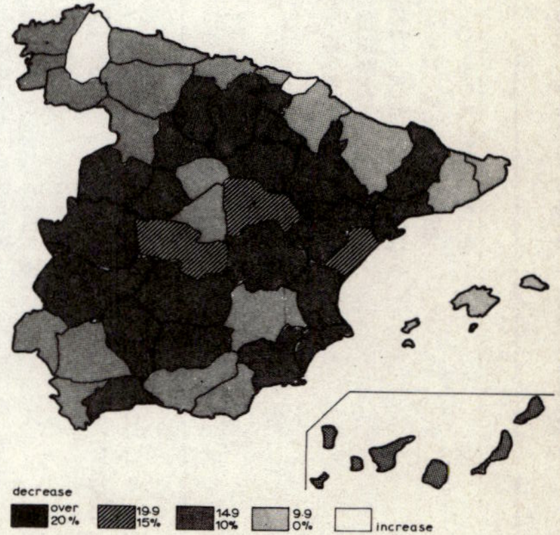

Sources: Banco de Bilbao, 1975

Table 8.8: Changes in Rural Population in Greece

	Percentage of agricultural population in total active population 1951	1971	Difference 1951-1971 %
1. Athens Region	1.9	0.7	- 1.2
2. Aetol-Akarnania	69.6	65.2	- 4.4
3. Boetia	67.3	45.1	-22.2
4. Euboea	55.6	38.8	-16.8
5. Evritania	79.3	68.2	-11.1
6. Phthiotis	63.0	51.2	-11.8
7. Phocis	67.5	48.2	-19.3
8. Argolis	67.9	35.4	-32.5
9. Ahaia	87.5	63.6	-23.9
10. Arkadia	53.3	25.2	-28.1
11. Ilia	69.8	61.2	- 8.6
12. Corinthia	65.2	38.1	-27.1
13. Laconia	72.8	69.2	- 3.6
14. Messenia	64.0	56.8	- 7.2
15. Ionian Islands	63.5	55.1	- 8.4
16. Arta	69.4	55.4	-14.0
17. Thesprotia	57.8	51.1	- 6.7
18. Yannina	55.1	45.1	-10.0
19. Preveza	62.8	51.0	-11.8
20. Karditza	74.3	62.1	-12.2
21. Larisa	61.3	44.2	-17.1
22. Magnesia	36.8	26.0	-10.8
23. Trikala	64.2	54.0	-10.2
24. Grevena	68.2	58.0	-10.2
25. Drama	72.6	53.0	-19.6
26. Imathia	61.7	51.0	-10.7
27. Thessaloniki	28.2	4.1	-24.1
28. Kavala	57.2	44.1	-13.1
29. Kastoria	55.7	35.1	-20.6
30. Kilkis	74.8	65.3	- 9.5
31. Kozani	67.3	51.2	-16.1
32. Pella	74.8	62.1	-12.7
33. Pieria	66.2	58.1	- 8.1
34. Serres	75.5	69.2	- 6.3
35. Florina	70.3	61.3	- 9.0
36. Khalkidiki	66.1	45.1	-21.0
37. Evros	69.9	61.7	- 8.2
38. Zanthi	75.4	65.3	-10.1

Table 8.8: Changes in Rural Population in Greece (continued)

	Percentage of agricultural population in total active population 1951	1971	Difference 1951-1971 %
39. Rodopi	70.9	70.8	0
40. Aegean Islands	55.4	35.1	-20.3
41. Iraklion	59.4	51.2	- 8.2
42. Lasithi	65.0	70.1	+ 5.1
43. Rethymnon	66.3	70.1	+ 3.8
44. Canea	52.7	51.2	- 1.5
GREECE	48.7	39.6	- 9.1

Sources: NSSG, 1951, 1961, 1971; Agricultural Statistics of Greece, 1963, 1969, 1974; Agricultural Bank of Greece, 1976; OECD, 1971, 1979; TEE, 1975

The highest percentage decreases are found in provinces around Madrid (e.g. Toledo and Guadalajara) and near the Mediterranean coast (e.g. Castellon), where peasants are searching for temporary jobs in the tourist sector (see Map 8.7). In Italy, the agricultural population in the south and in the centre is declining faster than in the north, but still remains high. Only three Italian regions have agricultural employment below 10 per cent, but, if provincial data are used, 31 provinces out of 91 have an agricultural population of below 10 per cent.[3] The geographical pattern in Italy also shows that faster declining regions are close to or include large metropolitan areas, such as Milan and Turin in the north, and Florence and Rome in the centre (see Map 8.5). In the south, out-migration still remains a predominant factor of agricultural population decline, while the attraction of industrial growth poles such as Bari-Brindisi-Taranto play a secondary role (INEA, 1976).

The overall reduction in the agricultural

Map 8.8: The Greek Province

1· Athens Region
2· Aetol-Akarnania
3· Beotia
4· Euboea
5· Evritania
6· Phthiotis
7· Phocis
8· Argolis
9· Ahaia
10· Arkadia
11· Ilia
12· Corinthia
13· Laconia
14· Messenia
15· Ionian Islands
16· Arta
17· Thesprotia
18· Yannina
19· Preveza
20· Karditsa
21· Larisa
22· Magnesia
23· Trikala
24· Grevena
25· Drama
26· Imathia
27· Thessaloniki
28· Kavala
29· Kastoria
30· Kilkis
31· Kozani
32· Pella
33· Pieria
34· Serres
35· Florina
36· Khalkidiki
37· Evros
38· Xanthi
39· Rodopi
40· Aegean Islands
41· Iraklion
42· Lasithi
43· Rethymnon
44· Canea

Map 8.9: Greece: Decrease of Agricultural Population, 1951-1971

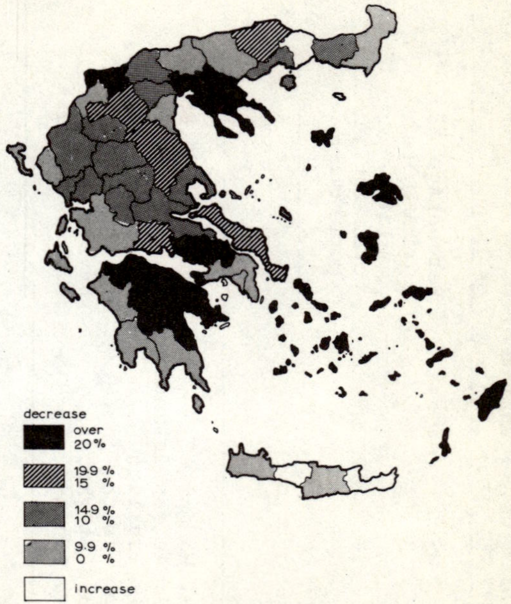

Sources: NSSG, 1951, 1961, 1971

population has been accompanied by far-reaching qualitative changes in its composition, which reflect the selective nature of most rural migration. In Greece over 48 per cent of all farm operators are over 55 years old, while in Italy the figure is just over 35 per cent. Change and modernization in farming are therefore difficult to be diffused with obvious implications for any development programme (Agricultural Bank of Greece, 1978). Another direct result of the rural exodus has been the increasing feminization of the agricultural work. In Spain, for example, the proportion of women among all employed in agriculture rose from 27 per cent in 1951 to 30 per cent in 1976 (OECD, 1977).

Farm Structure and Land Fragmentation

This played a key role in southern European agriculture, both in terms of size of farm and of land ownership (see Diagrams 8.1 - 8.3). One must take into account the important variety of farm structure among the three countries: Greece being on the one extreme with highest percentages in 1 to 5 hectare farms, and Spain on the other, with highest figures for farms above 100 hectares. Note also that size categories differ for each country, making direct comparisons somewhat difficult.

The most serious problem is the small size of farms, aggravated by considerable fragmentation and the wide disparities between one region and another. In Italy, farms of less than 5 hectares take up only 18.5 per cent of the total farmland, but represent over 75 per cent of the total number of farms. In Spain, the minifundia holdings below 5 hectares account for more than 50 per cent of the number of farms but only 6 per cent of the total agricultural area. Small farms are most prevalent in Greece: about half the total cultivated area consists of farms of less than 5 hectares, while farms of over 20 hectares account for only 7 per cent (OECD, 1975).

Nevertheless, the degree of polarization of landholdings in a few hands still is important, mainly in Spain and Italy. Greece, on the other hand, has no such pronounced polarization, as over 90 per cent of its utilized agricultural area is made up by farms of less than 20 hectares in size. This is the result of an early agrarian reform started in 1871 and ended in 1922 - something that the two other countries at that period were lacking.

Land fragmentation has its origin in deep historical and cultural tradition. For France and

Figure 8.1: ITALY, 1970 Figure 8.2: SPAIN, 1970 Figure 8.3: GREECE, 1971

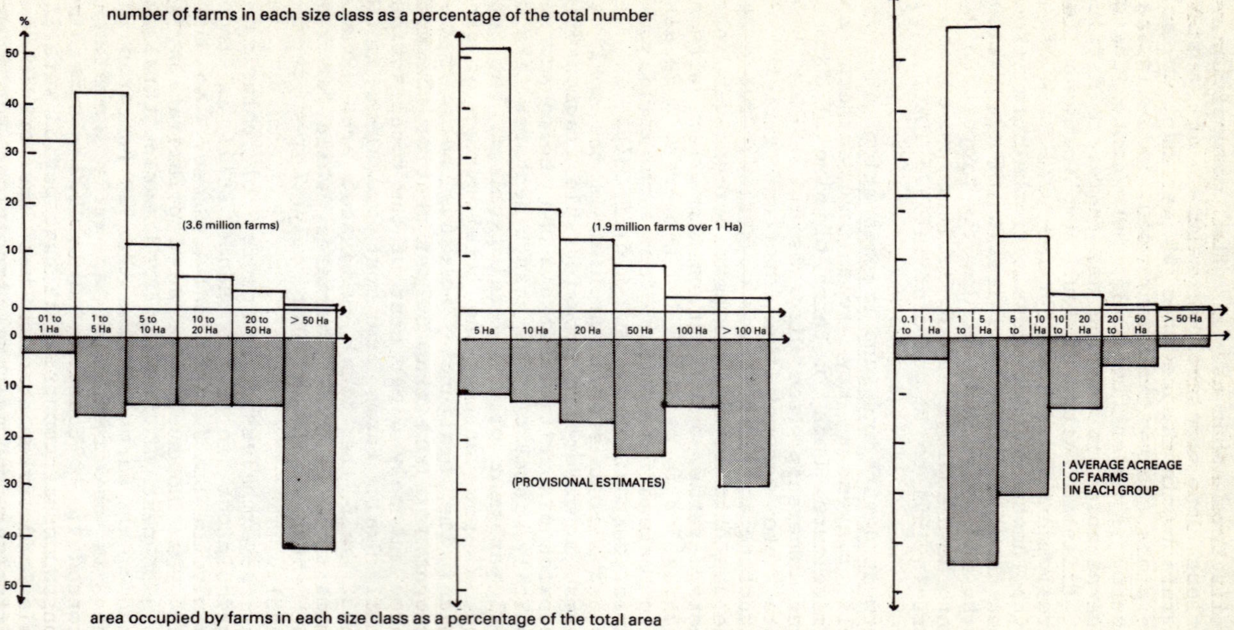

Source: OECD, 1975

the Iberian peninsula its origins can be traced back to the Napoleonic code of equal inheritance usually to sons. For southern Italy and Greece, in addition to this there still exists the social institution of the dowry, where daughters must have a house or a piece of farmland in order to get married (Jones, 1984; Skouteri-Didaskalou, 1984).

The problems raised by fragmentation are largely economic, though there are notable socio-psychological and administrative disadvantages. In the case of Greece, farms, small as they are, comprise seven plots on the average, which means that the average size of a plot is only 0.4 hectares (Greek Ministry of Agriculture, 1974). In Spain, there were 14 plots per holding on the average between 1962 and 1972; however, the number of plots fell from about 39 million to 27 million over the same period (OECD, 1974). In Italy, while farms of over 20 hectares account for more than half the total area, some of these farms are located in mountainous or hilly zones of low productivity and one must not be misled by the areal size as regards their productivity (Ministerio di Agricultura, 1975).

Fragmentation levels have important regional variations. They are particularly high in the Greek island communities (e.g. Crete and the Cyclades average 10.8 parcels per farm) and in the Peloponnese. In northern Spain, holdings on the average have more than thirty parcels, while in Italy fragmentation is more prevalent in high hill and mountain areas, both in the north and south. This pattern persists despite various land reform programmes aimed either at aggregating small units into larger ones or breaking up large units into smaller ones (e.g. Italy, 1950, in the Mezzogiorno), and despite various attempts at consolidation of small farms (as in northern Spain from 1952). While such attempts have been important for political reasons, overall they have done little to alter the basic farm size distribution of southern European agriculture (see Jones, 1984).

8.2 THE 'DONOR' ROLE OF AGRICULTURE IN SOUTHERN EUROPEAN CAPITALIST DEVELOPMENT: A DEBATE

The role of agriculture in southern Europe and its economic, political, cultural and spatial consequences have been interpreted in various ways. Using different terminology and in many cases different methodologies, many social scientists have viewed Mediterranean agriculture as a traditional and backward sector, based on 'archaic' and inefficient forms of organization (Aiello, 1972; Filias, 1972). According to these views, the alleged 'backwardness' of southern European peasants is due to their isolation, lack of technical modernization and ineffective integration into the competitive and standardized market, i.e. into the CMP. According to some national and EEC experts, southern European agriculture, for reasons internal to it, was neither able to generate savings nor to perform as a market for products and services produced in other sectors and regions.[4] Thus, for many agriculture contributed little to the development of capitalism in Italy, Spain and Greece, while the result of this inadequate performance was the preservation of a kind of dualistic dichotomy between the cities and the countryside.[5]

These observations and ideas create a set of general beliefs that peasants in southern Europe have a strong penchant for being idle and have low regard for thrift; that their lack of modernization does not permit a rational cultivation programme; that their low incomes are due largely to their inability to save and to adapt themselves to new techniques and new demands of the market; and finally, that these inadequacies and especially the dualistic dichotomy between urban and rural regions can be overcome only through deliberate and innovative planning policies.[6]

I am suggesting, to the contrary, that agricultural performance in southern Europe was neither inadequate nor that it has led to a dualistic urban-rural dichotomy. Most agricultural regions in southern Europe are not able to generate savings and to accumulate capital because of a number of specific sectoral conditions (heterogeneous agricultural activity, degree of mechanization, multiplicity of entrepreneurial forms, capitalist versus peasant farms), and labour market characteristics (land-ownership, manual labour, part-time farming, independent/dependent farmwork). Such an approach centres on inter-relationships, considering them above all in the light of class relations. These relations constitute the very mode of articulation of rural regions with the rest of social formation and the international market. The articulation has been developed historically around certain core-semiperiphery-periphery structures, where economic, political, cultural and spatial elements are combined together to subordinate certain forms of agricultural production and to transfer surplus from them to other production forms and to other sectors and regions.

This is indirectly acknowledged by some liberal economists and planners arguing about the monetary expression of surplus transfer as 'transferring resources, income and funds'. These transfers from agriculture to other sectors have been seen as a legitimate and necessary step towards economic progress in developing economies. Kuznets (1956) for example, argues about the 'factor' and 'market' contributions of agriculture to other sectors. A factor contribution occurs '... when there is a transfer or loan of resources from agriculture to another sector'; a market contribution occurs '... when part or all of the agriculture product is offered at low prices in exchange for goods produced by other sectors, at home or abroad ..'[7] Indeed, one of the crucial problems of capitalist growth is how to extract from the agricultural sector a surplus - in terms of both capital and labour - for the financing of accumulation in other sectors and regions, mainly for industrialization in urban centres (Griffin, 1979; Goodman and Redclift, 1981). This extraction of surplus Kuznets calls 'forced contribution' (see also, Johnson and Mellor, 1967) and its conventional justification runs as follows: capital formation (i.e. accumulation) depends on surplus (surplus value or surplus labour); the majority of people in developing regions and countries are in agriculture; hence, a major part of capital for development must come from the extraction of surplus in agriculture.

According to OECD (1979) the continuing subsidy to industries in Spain and Greece during the 1950s and 1960s became necessary simply to support the level of output and value added in such industries. Therefore the agricultural sector was being 'taxed' not to increase overall saving in the economy, but rather to give an ongoing subsidy to industries that were unable to compete in international markets, even with a correction for the over-valuation of currencies that existed in countries following this pattern of growth.

In Italy, Saraceno (1959) described the historical relationship between agriculture and industry and the mechanisms by which the former transferred funds and resources to the latter. For Saraceno, these transfers occur in three ways: through the agricultural surplus product; through the movements of factors of production (e.g. capital and labour in agriculture); and through the market (e.g. unfavourable prices for agro-products) aggravated by inflationary tendencies. Similarly, Kaldor (1975) argues that the problematic position of peasants has its roots in inherent advantages of 'processing' versus 'land based' activities which are basically due to economies of large-scale mechanized production.

Describing regional inequalities in Europe, Kaldor emphasizes the fact that trade between two regions, one specialized in land-based activities such as agriculture and the second in processing activities such as industry will mainly benefit the second, because for Kaldor:

> ... in industrial production, contractual costs form an important independent element in price-formation; competition is necessarily imperfect; the sellers are price-makers, rather than price-takers. Whereas in agricultural production incomes are derived from prices, in industrial production it is prices that are derived from, or dependent on, contractual incomes (i.e. on the level of wages). (N. Kaldor, 1975, p.14, emphasis added)

Kaldor's argument, which echoes Emmanuel's concept of unequal exchange in the narrow sense, parallels the thesis developed by Schickele (1968). Schickele argues that agriculture continues to be a 'blood donor' to the industrial and service sectors of the economy. Introducing a similar argument to that of Kaldor, he concludes that:

> ... savings from agriculture ... are already transferred elsewhere through the working of the land tenure system with its rent charges and taxes, the credit system with its interests and the market system allotting farmers a very weak bargaining position and often unduly low prices for their products. The almost universal income disparity between farm and non-farm population, resulting from these conditions, represents an income transfer in the sense that what farmers lose by a low level of food prices, city people gain by the corresponding reduction in their food expenditure which benefits industrialization at the expense of agriculture. (Schickele, 1968, p.7, emphasis added)

This income transfer to the benefit of industrialization has been legitimized as a rational allocation of resources; any deviation from this 'ideal' has been interpreted as a misallocation, a constraint to the 'natural course of economic development during which resources are transferred from agriculture to other uses' (Hill, 1984, p.90). This is also the foundation of neo-liberal criticism against the CAP. Critics argue that the CAP encourages all farmers to expand production in order to maximize their net receipts or minimize their net contributions. Through this inefficient policy, CAP transfers income from consumers to farmers. It follows that countries or regions with a larger share of consumers than farmers transfer income to countries and regions with a larger share of farmers, that is, net food importers transfer income, due to CAP, to net food exporters (see Hill, 1984).

It is true that CAP price support has maintained agricultural incomes at a higher level than would otherwise have prevailed. Also, the regional pricing policy of the CAP, as best exemplified by cereals, has not encouraged change in production patterns. But these observations cannot justify arguments of net income transfer to farmers without a consideration of what farmers have already contributed to national or EEC income, or without an analysis of important regional variations among farmers themselves. Thus, if Denmark, France and Holland benefit from the CAP together with Ireland, but not Italy or the UK (see Table 8.9), this is so because of the specific price policy of the CAP, the products it protects, and the regional differences in terms of factors of production among European farmers. CAP has taken agriculture within the EEC as a honogeneous entity, so that its price policy has almost surely widened existing regional inequalities. Thus, an answer to the misallocation of the CAP's resources is not to abolish the CAP altogether - as neo-liberals argue - but to reorient it and to incorporate it within the regional policy of the EEC (see Bowler, 1976; Papayannakis, 1978).

It is precisely this income transfer emphasized by the CAP critics as well as by Schickele, Kaldor and others (Sereni, 1956; Schultz, 1965) that a number of Marxists analysed from a different perspective, basing their investigation on class analysis and the role of the state. For liberals (and for some radicals, too)

Table 8.9: Estimated financial flows arising from CAP in £ millions

	1977 FEOGA and TRADE transfers	1978 FEOGA and TRADE transfers	1979 FEOGA and TRADE transfers
Germany	-824	-556	- 590
France	+589	+616	+ 855
Italy	-569	-932	- 696
Holland	+800	+846	+ 654
Belgium/ Luxembourg	- 24	- 21	- 251
UK	-637	-783	-1,107
Ireland	+402	+527	+ 614
Denmark	+542	+683	+ 710

Notes: FEOGA: (European Agricultural Guarantee and Guidance Fund)
FEOGA transfers: financing prices, improvement of marketing, structural changes and other budgetary transfers.
TRADE transfers: transfers from the consumers of one country to the producers of another via higher prices due to CAP.

Source: Hill (1984)

whenever the unequal nature of this articulation is admitted, it is seen as a conflict between the countryside and the city or between agriculture and industry rather than between classes. Consequently, this approach has been ideologically useful for those who, for political reasons, have attempted to minimize awareness of the development of both internal class differentiation and socio-spatial fragmentation in the countryside.

It is necessary, therefore, to reject the over-simplified and misleading concepts of archaic and inefficient forms of agriculture in southern Europe and the dualistic dichotomy between urban and rural regions. The notion of a 'negative' articulation of agriculture with other sectors can be retained, however, although it requires a totally different explanation. Behind what liberals call 'income transfer' there exists a set of powerful class relations and other forms of domination that manifest themselves in numerous ways. These relations are not manifest only in circulation but also in production relations. In this interpretation, one thing is certain. No phenomenon or process relevant to the study of the performance of agriculture (whether it be growing mechanization; the marketing of products; the effects of various measures of state intervention; composition of the agricultural labour force; property relations, etc.) can be adequately investigated if it continues to restrict itself to the agricultural sector alone, or to its aggregate economic performance (Mouzelis, 1978; De Janvry, 1981; Mottura, 1980), if, in other words, it does not bring out the specific mode or modes of articulation between various forms of agricultural production, their relations with the rest of the capitalist economy and the state, and the practices of social agents within civil society.

8.3 THE FORM OF PETTY COMMODITY PRODUCTION IN AGRICULTURE AND THE IMPORTANCE OF SMALL PEASANT PRODUCERS

From a Marxist perspective, the role of agriculture in southern European capitalist development is considered to be of key importance. Various Marxists have emphasized the existence of a 'network of exploitation' between urban and rural regions. This network of exploitation manifests itself in various ways. As Emmanuel (1969) and Vergopoulos (1975) argue, the low cost of labour in southern European agriculture has maintained the prices of basic agricultural commodities at an artificially low level, and this in turn has made it possible to pay the urban working class lower wages, since their cost of living was largely tied to the prices of these agricultural goods. Amin (1977) has gone further, arguing that the prices and wages in agriculture that help to maintain the low wages in urban centres were beneficial for foreign capital and the local oligarchy, but not for the CMP as a whole, because these low wage levels in agriculture and industry negatively influenced the development of the internal market (national and regional), which in the last analysis means a restricted realization of surplus value.

Building upon these observations, a good deal

of controversy has arisen among Marxists in southern Europe over the question of why capitalist development has not transformed the agricultural sector to a greater extent. Many different attempts have been made to explain the continued importance of what appear to be pre-capitalist modes or forms of production and, in particular, the large number of peasant producers with smallholdings operating on a family basis.[8] Within the broad spectrum of analysis and debate, however, two opposing points of view can be identified.

First, we have those who, following the traditional Leninist view, claim that there is no specific peasant mode of production in southern Europe (Boccara, 1971; Panitsidis, 1983; for Latin America, see De Janvry, 1981). Peasants are merely an ephemeral class or an unstable class fraction that are destined to disappear with the full development of capitalism. They are - in the Leninist terminology - a transitory fraction, being absorbed by the essential classes: bourgeoisie and proletariat. Contesting this view are the so-called 'neo-populists', who claim that peasants do constitute a specific mode of production, defined as petty commodity production, that is articulated with and dominated by other modes (Servolin, 1972; Amin, 1974; Vergopoulos, 1975; Mouzelis, 1978; Psichogios, 1984).

Starting with the first approach, both Lenin and Kautsky analysed agriculture without postulating a specific peasant mode of production. Marx, however, did frequently use the term 'peasant mode of production' but he used it - according to this view - to characterize the technological and social organization of the labour process in agriculture (De Janvry, 1981). According to Panitsidis (1983) the peasantry has a system of producing which isolates each peasant from others, but this is not sufficient grounds to distinguish a separate mode of production. The rural proletarian very often has the appearance of a peasant, owing to his control over a small plot of land, but the social relations that characterize his labour process are more those of a worker. From this it follows that the peasantry is almost by definition polarized on the working class side. This was a controversial issue in the French Communist Party during the 1960s and 1970s, when the party followed contradictory policies favouring both the small peasant producer, as being proletarianized, and the development of big state-owned farms with organized farmworkers.[9]

Those who follow this approach tend to deny the persistence and development of the simple or petty commodity mode of production in the countryside, taking it as a purely transitory form between feudalism and capitalism (Boccara, 1971). Its theoretical clarity, however, seems superfluous for the analysis of existing social contradictions in the southern European countryside. It falls into a technological determinism, where the development of the so-called 'productive basis' will determine the sequence of appearing forms of social organization (see also, Dedousopoulos, 1985). Local class organization is today much more complicated than the functionalistic polarization between two classes only: bourgeoisie and rural proletariat. Such an approach misses entirely important internal class differentiations and issues such as discontinuous and spatially differentiated labour markets, part-time farming, irregular employment in industry and tourism. It is true - and here I am in agreement with this view - that a peasant <u>mode</u> of production does not exist, since capitalist social relations are dominant all over southern Europe. What this approach is missing, however, is the existence of a different <u>form of production</u>, which expresses specific social relations (share-cropping, different types of tenant farming, part-time employment in other sectors, ownership of means of production, family-based production units, small-scale operations), that are not peculiar to agriculture but can be extended to describe small-scale household activities in the so-called 'informal sector', in industry and services' activities as well (Pugliese, 1982). In this case we must distinguish the category of mode of production from the category of form of production, and we must speak of the <u>articulation of various forms of production</u> within the CMP. Their rank and importance are organized politically, that is, through various forms of struggle among social agents in the countryside. It is therefore wrong to argue that peasants and simple commodity production are disappearing social categories, when what is actually happening is a <u>transformation</u> of their roles and positions within the wider system of social reproduction.

On the other side of the spectrum there are those who advance the theory of petty commodity production in the small peasant economy, negatively articulated and subordinated to the capitalist mode of production. Capitalism, according to this second approach, <u>needs</u> such a mode of production because it can easily exploit peasants and transfer surplus for accelerated accumulation in other sectors. Thus, the persistence and development of petty commodity production is conceived as a functional necessity for the accumulation process as a whole. For southern Europe, the work of Servolin (1972), writing about France and Italy, has been highly influential. Servolin (1972) argues that the primary target of the small peasant producer is not the valorization of his capital but the simple survival of the producer and his family. Attaining an average rate of profit is not a necessary condition for their operation.[10] According to Servolin, agricultural work is continuously intensified and extends over longer hours. What remains from their working effort (minus expenses, taxes, loan payments, etc.) is the equivalent of a <u>wage</u>. In fact, small peasant producers in certain southern European regions are becoming <u>wage-earners working by the piece</u>, what Amin calls, 'home-staying proletarians'. In his words:

> ... on the surface, the peasant remained a commodity producer who offered products on the market but in actual fact was a seller of labour-power, this sale being masked under the cover of petty commodity production. (S. Amin, 1974, p.60)

In this respect, land ownership or land control (in the case of share-cropping) in small agro-businesses could be seen not as a privilege but as a condition for self-exploitation in the agricultural sector.

Vergopoulos (1975) writing about Greece similarly conceives peasants as members of a simple commodity mode which relies both on the organization of the peasant household, where only family labour is used, and on the objective of production for subsistence versus profit. As he specified for Greece, if capitalist accounting methods are applied, the majority of Greek peasant 'enterprises' (a term defined by law and used by the National Statistical Service) operates at a deficit. The Greek state lies in a position to substitute production in these small peasant 'enterprises' for imported goods. The same is true for Spain and Italy (OECD, 1975).[11] Indeed, indebtedness and the need for money-income to purchase certain commodities (e.g. tools and medicine) require many small peasant producers to seek other wage-earning employment in order to supplement their inadequate income. The rural masses tend to be forced into a dual status of independent producers and 'dependent' farm-workers, for agro-industries, big land-owners or other small peasant producers.

Vergoupolos and Amin's thesis has been criticised by Mouzelis (1978). Mouzelis, while he partially accepts simple commodity production in southern Europe, claims that low agricultural incomes are due to lower agricultural productivity and not to supposed negative articulation of the agricultural sector with the rest of the economy. He supports his argument by providing evidence from other sectors (industry and tourism) in relation to productivity figures from other European countries. Mouzelis rejects the negative articulation thesis as the major means for exploiting the peasant economy and concludes that private capital and the state in Greece (and in other southern European countries) prefer to exploit the direct producer in the sphere of circulation through the market mechanism.

A similar view is developed by Dedousopoulos (1985) who criticizes the functionalism of Vergopoulos and Amin's thesis. Capitalism, he argues, is not a self-regulated system, capable of identifying its 'needs', so that petty commodity production performs its assigned roles. The persistence of small peasants in southern Europe (along with big capitalist farms) must be seen as the outcome of specific social and political conditions, of class struggle and the mode of political domination in the countryside. Dedousopoulos supports the argument that the dominant social relations of production in petty commodity production give a special importance to the sphere of circulation. Thus, commercial capital and the state play key roles in the extraction of surplus from petty commodity producers and in its distribution to other sectors and individual social agents.

The critique made by Mouzelis and Dedousopoulos reintroduces the concept of petty commodity production, without, however, falling into functionalistic and deterministic explanations. Following a similar logic I will try to formulate an alternative view of peasants in southern Europe. As in the case of the related debate between autonomous and surplus transfer theories of development (see Chapter 3), the formulation of an alternative view requires a critical but not dogmatic position.

Central to such a critical alternative interpretation of the agrarian question in southern Europe are the following observations which draw upon the arguments of this and earlier chapters.

1. Capitalist development in agriculture must be defined in terms of class relations and other forms of social and political dominations which are not restricted to a simple polarization between bourgeoisie and rural proletariat. It requires an analysis of a multiplicity of forms of production in the countryside and the tendencies towards equalization/ differentiation of such patterns as land-owernship, labour process, political mediation and ideological/cultural domination.
2. Regional patterns of extremely heterogeneous physical and technical conditions of production must be taken into account. These 'background' patterns are insufficient in themselves and are ideologically loaded as explanations of regional variations in agricultural performance. But we should avoid a reductionist vision which does not see them as initial factors of disequilibrium, whose cumulative effects have been exploited further by certain class alliances to produce socially differentiated regional structures.
3. Internal and external factors influencing the development of capitalism in southern Europe are dialectially interrelated, and they require an emphasis on the continuing importance of historical and geographical conditions. The analytical approach which I propose focuses on the analysis of forms of agricultural work and types of farm enterprises associated respectively with local social classes and their position in the social and spatial division of labour. A typology of working conditions and farm enterprises is therefore necessary in that it will permit us to identify the parallel existence and development of big capitalist farms and petty commodity producers. The historical and geographical analysis will also permit us to identify the channels of surplus extraction which are different for every specific form of production.
4. Petty commodity production is differentiated from CMP in five major aspects. First, production takes place in a large number of

small units, where the producer is also the owner of his means of production. Secondly, the identification of the same person as the owner of production and the worker, leads to a search of exploitation in the sphere of circulation. Thirdly, the petty commodity producer is a social entity which includes all family members capable of work. Fourthly, 'wages' and 'profits' in the 'pure' capitalist sense do not exist. Family members are rewarded for their labour through kinship relations, while the family as a whole is taking money through the market or incentives from the state. And fifthly, commercial capital and the state play a much greater role in influencing production than in capitalist relations.

5. The law of value is not valid in petty commodity production. What is extracted is surplus labour and not surplus value. It is a surplus which is largely influenced by non-economic reasons and the balance of political forces in a social formation. All this diversity may be placed under two fundamental types of transfer which, until further specified, we shall term intra-sectoral and inter-sectoral transfers. As I will try to show in Chapter 10, a number of localized conditions in the sphere of production and circulation combined with relatively high agricultural productivity permit in certain peripheral regions higher rates of surplus labour. In many cases, however, there is a conjunction of large surplus extraction with high rates of agricultural growth, but usually with low levels of living conditions and inadequate means of collective consumption in peripheral agricultural regions.
6. In the period studied in detail (i.e. 1950-1975) the articulation of various forms of agricultural production in southern Europe with the CMP sustained accelerated (if unstable and unequal) accumulation in other sectors and regions with opposite effects for the agricultural sector. In this respect, the corresponding pattern of uneven geographical development contributed to capitalist accumulation in a period of widening socio-spatial inequalities. This development did not destroy small peasant producers: on the contrary it has recently transformed other more centralized and tailorized working processes to

diffused rural household sites engaging in activities which involve not only agricultural production but also industry, tourism and services.

NOTES

[1] 'Withdrawal' price is the price at which producer groups can withdraw certain fruits and vegetables from the market. It is a much debated issue in EEC policies and a most favourable one among the peasants. It was also part of Greek national policy before joining the EEC.

[2] The study used factor analysis to come out with these groupings. The various regional groups of crops in the three countries have percentages larger than national averages and cover in each case more than 40 per cent of ASU in each region.

[3] See Annuario de Estadistica Agraria (1977).

[4] See OECD, 1975, 1977 and EEC, Agricultural Committee 1976. For a critical overview of this approach, see Vergopoulos, 1975, and Papayannakis, 1976.

[5] The concept of dualism has been widely discussed, especially for Italy by Chenery (1968) and Lutz (1962) and criticized by Holland (1971). An alternate interpretation from a Marxist perspective has been provided by Mingione (1978), with his concept of 'neo-dualism' based on differences in regional occupational structures.

[6] See the programmes of agricultural reconstruction in the 1950s and 1960s in Greece, Spain and Italy: McNeill, 1978; World Bank Report, 1966; Saraceno, 1959.

[7] A view similar to that of Kuznets has been developed by Preobrazhensky concerning agricultural development in the Soviet Union. According to him, economic resources have been transferred from agriculture to industry through (a) an increase of industrial product prices and a decrease of prices in agriculture; (b) a relatively higher taxation of the agriculture sector; (c) a policy of forced public loans; (d) a policy of inflation which redistributes resources from agriculture to industry.

[8] For a review of the debate, see Vergopoulos (1975) and Lipietz (1977), Chapter 2, Section III, pp.36-53. Lipietz uses two maps from Gervais, Servolin and Weil showing the decline of peasant population in various French provinces.

[9] PCF still remains weak in its agricultural

policy, promoting contradictory and narrowly nationalistic theses against the EEC and against Greece's, Spain's and Portugal's entries to the EEC.

[10] This is not true for all regions and provinces, the spatial differentiation being one major source for DLA and GTV.

[11] The truth of the matter is that the prices of some products are kept very high under the pretext of protecting the poor peasants on the worst land. One example of this practice has been the maintenance of the price of wheat in Spain and Greece above the world market price, which has allowed its production to substitute for imported wheat. See OECD, 1974.

Chapter 9

ACCUMULATION IN AGRICULTURE AND REGIONAL DIFFERENTIATION

We have seen so far that the political, social and economic developments since World War II in southern Europe have transformed Mediterranean agriculture almost entirely into an induced sector, dependent for its growth on the rest of the economy and undergoing deep transformations provoked by causes exogenous to agriculture itself. With the expanded development of capitalism, the traditional relationship was thus inverted: economic growth, instead of being an effect, became the 'cause' of agricultural growth.

The transformation of Italian, Spanish and Greek agriculture has been studied from a sectoral perspective at the national scale (Saraceno, 1959; Van Nieuwenhuije (1972: OECD, 1971, 1979; EEC, 1972, 1976). Some of these studies have provided evidence of the transfer of both capital and labour from agriculture to other economic sectors following the process of capitalist growth and modernization (see also, Schickele, 1968; Kaldor, 1975). While there is general agreement about this 'donor' role of agriculture in capitalist development, little attention has been paid to the regional consequences of this process and to the role of agriculture in uneven regional development.[1] The prevailing emphasis on industrialization, growth poles and tourist development has kept the role of agriculture in the shadow, as a necessary but unimportant sector in regional development. Even in the case of the Cassa per il Mezzogiorno, whose initial targets in agriculture have already been described, little attempt was made to connect agricultural and regional development. The emphasis was, as Cottone (1972) points out, in integrating southern rural provinces into the advanced industrial north, mainly through sectoral policies at the national level. I argue, however, that an approach to the differential role of regions in the capitalist transformation and incorporation of agriculture provides a useful view of the process of uneven regional development operating today in southern Europe.

This chapter attempts a comparative empirical analysis aimed at illustrating the pattern of GTV in the agricultural sector in Spain, Italy and Greece. The analysis is based on a set of comparative official data available for the three countries and from the period 1950-1975. This periodization is particularly significant for southern Europe, because it coincides with the initial stages of rapid industrialization and general economic growth which was largely supported by a 'squeeze' of surplus at the expense of agricultural population and regions. 1975 was also a major political turning point, after which the authoritarian mode of social regulation necessary for the 'squeeze' of agricultural surplus began to disintegrate.

9.1 REGIONAL ACCUMULATION TRENDS AND THE AGRICULTURAL SECTOR

A necessary starting point for investigating the role of the agricultural sector in the regional development of southern Europe is to analyse the pattern of Differentially Localized Accumulation (DLA) in various regions. The DLA for the agricultural sector is dependent on the current performance of the various agricultural productive units located in a particular region and on the incomes that they obtain. This in turn relates to the performance of all sectors and to the total regional income. An analysis of the regional performance as a whole, however, without breaking it down to social and spatial aggregates (e.g. social classes, sectors and smaller spatial units) may generate problems in terms of <u>intra</u>-regional differentiation, which in many cases could be more important than <u>inter</u>-regional differentiation. In this respect, my discussion of the agricultural sector from a regional point of view must be seen as an example of suggested methodology and as one among many possible indications of uneven regional development in southern Europe.

The availability of information and data to construct a complete picture of DLA on a regional scale is limited. For such a picture to be drawn systematically, special data would have to be collected, a task far beyond the scope of this study. Thus, taking into account the data constraints and the available statistical information for the three countries, a first approximation of DLA could be achieved through a comparison between gross agricultural income (GAI) and agricultural gross value added (AGVA), on a regional and/or provincial basis. Gross agricultural income includes the returns to both capital and labour employed in agriculture, plus income derived from special subsidies, such as incentives to capital and social security benefits to labour.[2] The regional distribution of gross value added in agriculture expresses the productivity and the economic weight of the different regions in each country and shows the differences in volume of agricultural production.

As Aglietta (1976) argues, the distribution of the total gross income of a sector and of a region into wages and profits is the monetary expression of the antagonistic and exploitative relationship between labour and capital, at the point of production, consumption and reproduction. Value added is in turn the monetary expression of the quantity of abstract labour expended in a specific production process and expresses the productivity of work, that is to say the combination of living labour with constant capital for each work post. Total value added (or total national income) is the monetary expression of social labour power that creates value in an economy (Liossatos, 1979). Thus, an agricultral sector or a region specialized in agriculture realizes an income equal to its respective contribution to national agricultural income in monetary terms only if sectoral or regional income equals sectoral or regional value added.[3]

While in a national economy as a whole every sector shows a balance between gross value added and gross domestic income, this is not likely to be the case on the regional scale. Because the various regions have different productive structures, they perform with different conditions of productivity and have different opportunities for realization of surplus value. The only possibility for all sectors and regions of the economy to yield positive profits is in a state of equilibrium prices, that must satisfy all three following conditions (see also, Liossatos, 1979).

1. The net national/regional product obtained at the end of a production cycle is distributed to

the agents of production only through the market; capitalists and workers are 'rewarded' in proportion to capital advanced in the beginning of, and labour performed during, that cycle.

2. Conditions of full capitalist competition and mobility prevail, that is capital can flow freely from one sector/region to another in response to more profitable opportunities.
3. Labour can also move freely from one sector/region to another in response to any wage differentials.

Such a state can exist if, and only if, the profit and wage rates - to speak for only two important variables - are uniform throughout the economy. It is not necessary to repeat again the arguments of Part I to support my thesis why such a situation can never appear. The contradiction of equalization/differentiation, the role of the state, capitalist competition itself and above all the struggle between capital and labour regenerate the state of disequilibrium which is the <u>modus vivendi</u> of capitalist development.

In a state of disequilibrium, let us suppose that the agricultural sector in a region A, creates a unit surplus value S_A, but actually receives a unit-profit P_A. The sector's contribution to total surplus value of the regional economy could be presented as S_AX_A (where X is a positive column vector representing gross regional output). Its share in total profits is P_AX_A (Morishima, 1979; Lipietz, 1981). Taking into account in the economy as a whole

$$S_X = P_X$$

if $P_A>S_A$, or $P_A<S_A$, or $P_A=S_A$, this signifies respectively that the agricultural sector in Region A receives more, less or the same profits of its contribution to the economy's total profits. This can occur only at the expense of at least one other sector or region. The difference between P_A and S_A - in our case the difference between GAI and AGVA - is a monetary approximation of the GTV taking place among the regions of a country. Thus, while prices and sectoral and regional profits may differ from values and sectoral and regional surplus values respectively, certain totals should remain invariant under the transformation of values into prices of production and then into prices in the market (Liossatos, 1979). In particular, this transformation should not yield total profits higher

or lower than total surplus value assuming a closed economy.[4] It can only 'redistribute' total surplus value among sectors and regions, as direct and indirect GTV. More generally, a region 'gains', 'breaks even', or 'loses' depending on whether the difference between GAI and AGVA makes its differentially localized accumulation (DLA) positive, zero or negative, respectively.

Similar evidence in terms of value-losing and value-gaining sectors and regions is presented by Marelli (1983). He introduces the concept of 'relative sectoral gain' defined as the total sectoral gain as a proportion of total value created in the economy, assuming also that the nation as a whole is a closed economy. In addition, Marelli identifies the 'relative regional gain of surplus value' as total regional gain out of total value created in the region. To obtain empirical estimates for Italy, he uses standard input-output tables, through which he estimates fundamental Marxian parameters, namely the value of labour power and the rate of surplus value. In Marelli's analysis the relation between the net gain of surplus value and the level of regional development is clear cut. The gainers in all sectors are the more developed regions (Piemonte, Lombardia, Liguria), while the losers correspond more or less to the less developed regions (Molise, Puglia, Basilicata, Calabria).

From the agricultural point of view, a first indication of regional DLA can be achieved when we subtract regional AGVA from regional GAI. Both AGVA and GAI will be presented as percentages of total national figures in money terms. In this subtraction a net figure close to zero means that the agricultural sector in this particular region managed a balance between what it has produced and what it has received for it. Similarly a minus or a positive sign will mean that the agricultural sector is receiving less or more respectively in comparison to what it has produced, i.e. to its contribution to the total national agricultural product. Putting these observations in the framework of the previous theoretical discussion, a positive sign means that the agricultural sector in this particular region realizes not only its own produced surplus value but also a portion of value added generated elsewhere. In this manner, the monetary expression of the difference between GAI and AGVA among the various regions of the three countries can be seen as a first approximation of the magnitude of GTV taking place within each country.

Marelli (1983) also uses percentages of total regional value in all sectors created in the economy. His computed transfers of surplus value are also approximate quantities, which nevertheless take into account the effects of sectoral differences in the composition of capital. In any case he agreed with Liossatos (1979) that it is not properly unequal exchange - in Emmanuel's sense - which gives rise to these transfers but a more complicated process. For the Italian case he concludes that 'the reason for the derived pattern of interregional transfers is obviously that the less developed regions (particularly the southern ones) are more specialized in sectors which lose surplus value: agriculture, construction, trade, etc.' (p.68)

In terms of the proposed methodology, a few clarifications are necessary. First as I mentioned earlier, AGVA and GAI will be presented as percentages of total national figures in money terms. This satisfies both our comparative purposes (e.g. different currencies in the three countries) and the theoretical assumption described earlier that within a country certain totals in terms of value added and gross income should remain invariant. In cases, however, where an illustration of absolute gains and losses is necessary, absolute figures will be presented as well. Secondly, differences between GAI and AGVA represent all types of transfers in the agricultural sector (i.e. intra- and inter-sectoral, direct and indirect GTV, etc.). This assumption has both positive and negative implications. On the positive side rests the fact that the proposed methodology encompasses almost all possible types of transfers. On the negative side this approach cannot specify either the exact magnitude and importance of each type of transfer, or the exact regional origin/destination of these transfers. All that we know in the end of our analysis is that some regions 'gain' while others are 'losing' and this is interrelated with GTV. Moreover, our data cannot identify problems of local devaluation, i.e. cases in which GTV does not occur, but the region 'loses' part or all of its produced value.

Tables 9.1-9.3 and Maps 9.1-9.5 show the difference between gross regional income and gross value added for the three countries. For Spain and Italy, in addition to data on the agricultural sector, data on the performance of all sectors are presented. For Greece, comparable data on all sectors for the period under study were lacking.

Table 9.1: Regional Differences between Gross Income and Gross Value Added in Italy

	Percentage (Gross Income) - Percentage (Gross Value Added)					
	1955		1965		1975	
	(a)	(b)	(a)	(b)	(a)	(b)
1. Piemonte	.65	1.80	.80	2.10	.70	2.80
2. V. d'Aosta	.16	- .15	.18	0.00	.15	.10
3. Lombardia	- .10	2.00	.31	3.80	.41	3.95
4. Liguria	- .15	.60	.21	.30	0.00	.40
5. Trentino	- .15	.20	0.00	- .30	.16	.10
6. Veneto	.41	.80	.31	.60	0.00	- .15
7. Friuli	0.00	0.00	.15	- .10	.16	- .40
8. Emil. Romagna	.85	1.10	1.15	2.00	.95	2.25
9. Toscana	.31	.10	0.00	- .40	- .35	- .30
10. Umbria	.15	.10	0.00	.20	- .12	.15
11. Marche	.35	0.00	- .55	-1.20	- .85	- .90
12. Lazio	0.00	.50	.35	.70	.31	1.10
13. Abruzzi	- .61	.30	- .35	- .20	.12	- .10
14. Molize	- .69	- .80	- .55	- .20	- .41	- .15
15. Campania	.28	- .95	- .31	- .50	- .21	- .60
16. Puglia	- .85	- .15	.10	.15	.12	- .40
17. Basilicata	- .32	- .50	0.00	- .40	.16	- .60
18. Calabria	0.00	- .90	- .25	-1.20	- .18	- .85
19. Sicilia	-1.12	- .80	- .45	-1.10	- .21	.10
20. Sardegna	- .40	0.00	- .55	- .90	- .31	- .15

(a) = agriculture
(b) = all sectors

Sources: INEA, 1955, 1965, 1975; OECD, 1975; Annuario di Statistiche Provinciali, 1953, 1963, 1975; Eurostat, 1970, 1979

A final remark considers the term 'region' and its socio-spatial status. Regional boundaries were chosen on the basis of statistical information available for each country.[5] Thus, in the following analysis I use predefined regions corresponding to available data for description of the existing situation only. An alternative regionalization

Map 9.1: ITALY: The Geographical Distribution of Receiving, Donor and Balanced Regions in the Agricultural Sector (GAI - AGVA)

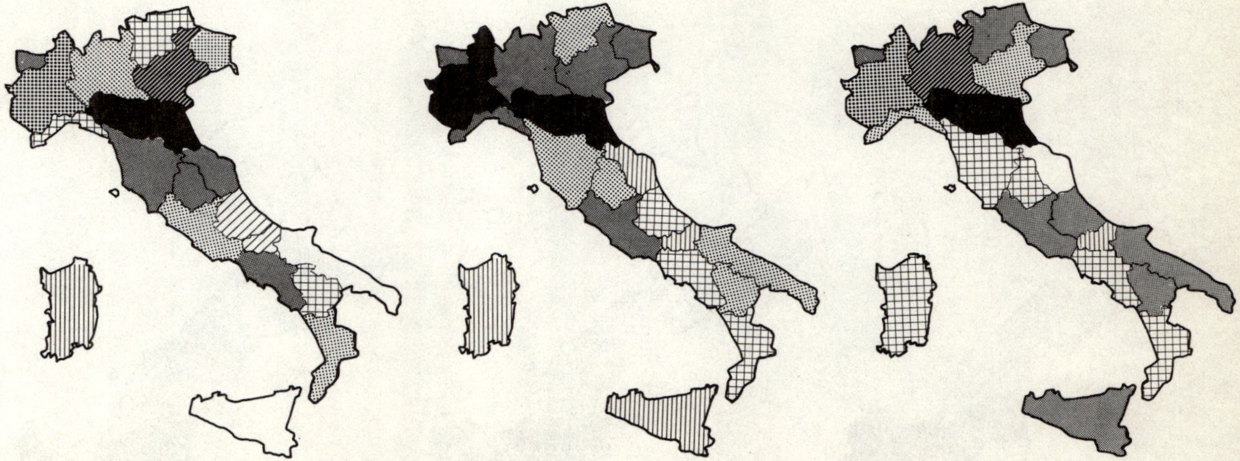

sources: INEA, 1955, 1965, 1975
OECD, 1975
Annuario di Statistiche Provinciali, 1953, 1963, 1975
Eurostat, 1970, 1979

Map 9.2: ITALY: The Geographical Distribution of Receiving, Donor and Balanced Regions Across All Sectors

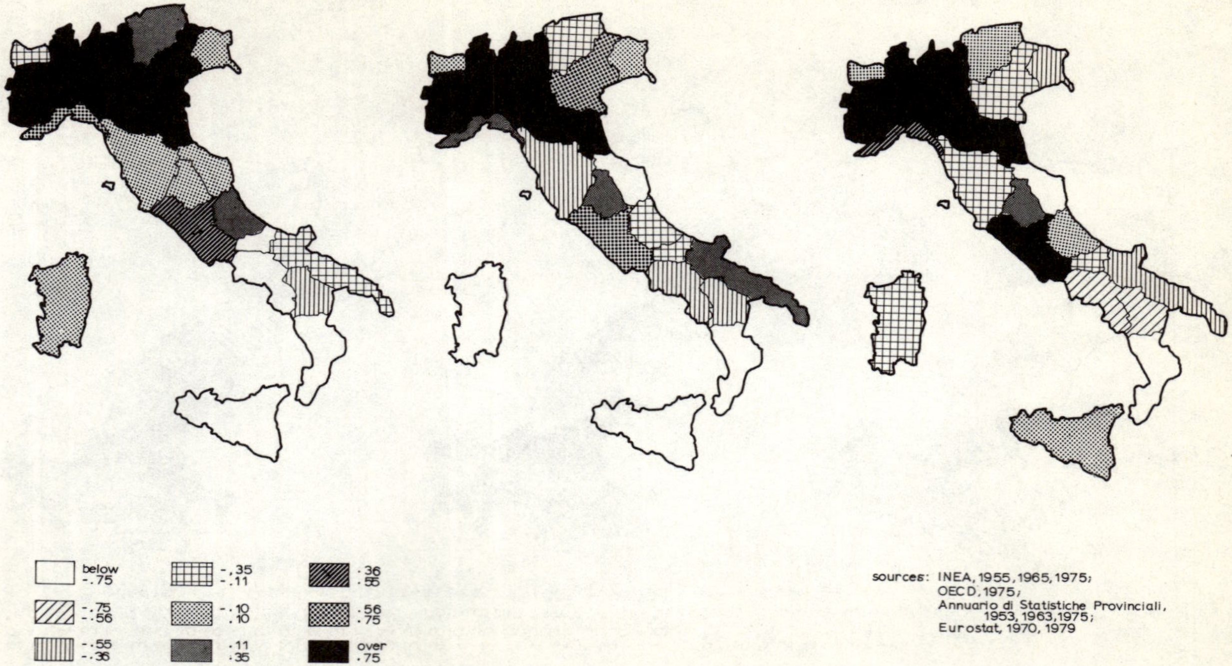

sources: INEA, 1955, 1965, 1975;
OECD, 1975;
Annuario di Statistiche Provinciali, 1953, 1963, 1975;
Eurostat, 1970, 1979

Table 9.2: Regional Differences between Gross Income and Gross Value Added in Spain

	Percentage (Gross Income) - Percentage (Gross Value Added)					
	1955		1964		1975	
	(a)	(b)	(a)	(b)	(a)	(b)
1. Alava	.13	.03	.13	- .03	.33	- .18
2. Albacete	0.00	- .02	0.00	- .05	- .19	- .22
3. Alicante	.25	- .11	- .56	- .18	- .34	- .36
4. Almeria	0.00	0.00	- .18	.01	.21	- .08
5. Avila	.10	- .01	0.00	.03	.12	.19
6. Badajoz	- .47	- .02	- .03	.31	- .39	.15
7. Baleares	.35	- .03	0.00	- .15	- .29	- .76
8. Barcelona	.40	- .76	.70	- .77	0.00	3.15
9. Burgos	.32	.10	.59	- .03	.82	- .21
10. Caceres	- .36	- .10	- .34	- .01	- .19	.09
11. Cadiz	-1.19	.23	-1.12	.20	-1.19	- .31
12. Castellon	0.00	- .13	-.00	- .12	- .19	- .33
13. Ciudad Real	- .41	- .05	- .49	.02	0.00	- .39
14. Cordoba	-1.32	0.00	- .89	.05	- .54	.10
15. Coruna (la)	0.00	.21	0.00	.18	.30	.17
16. Cuenca	.20	- .03	0.00	- .05	0.00	- .14
17. Gerona	.13	- .07	.10	- .29	0.00	- .21
18. Granada	.13	.18	.10	.26	0.00	.27
19. Guadalajara	.21	- .06	.12	0.00	.19	- .15
20. Guipozcoa	- .81	- .58	- .52	- .06	- .18	.22
21. Huelva	- .91	- .43	- .68	- .49	-2.27	-2.80
22. Huesca	.21	- .13	.42	- .41	.23	.31
23. Jaen	- .31	.38	- .21	.30	- .72	.21
24. Leon	.62	.14	.56	- .35	1.21	.05
25. Lerida	.33	- .24	.50	- .21	.84	- .30
26. Logrono	.86	- .08	.47	- .11	.52	- .20
27. Lugo	.70	- .01	- .30	- .10	- .83	- .09
28. Madrid	.10	1.15	1.50	2.18	.53	6.20
29. Malaga	- .74	.17	- .94	0.00	- .66	.21
30. Murcia	- .26	- .07	- .36	.02	- .69	- .95
31. Navarra	.69	- .08	.27	0.00	.63	- .15
32. Orense	- .38	- .08	- .52	- .21	- .64	.11
33. Oviedo	.17	- .02	.40	- .59	- .91	- .94
34. Palencia	- .11	- .10	0.00	- .18	.50	- .21
35. Palmas (las)	- .20	.04	- .12	- .31	- .38	.22
36. Pontevedra	- .61	- .09	- .20	- .04	- .50	.29

Table 9.2: Regional Differences between Gross Income and Gross Value Added in Spain (continued)

	Percentage (Gross Income) - Percentage (Gross Value Added)					
	1955		1964		1975	
	(a)	(b)	(a)	(b)	(a)	(b)
37. Salamanca	.16	- .01	.23	- .32	.26	- .20
38. S.C. Tenerife	0.00	.01	- .10	- .41	- .12	- .23
39. Santander	.84	- .09	.28	- .58	.50	.10
40. Segovia	.35	- .03	.11	- .08	.48	.08
41. Sevilla	- .55	- .08	- .30	.58	-1.04	- .61
42. Soria	.41	.01	.12	.01	.25	- .09
43. Tarragona	.10	- .17	0.00	- .18	- .24	-1.10
44. Teruel	.22	- .01	.18	- .09	.44	- .16
45. Toledo	- .37	0.00	- .47	- .10	- .21	- .29
46. Valencia	- .10	- .06	.65	- .02	- .93	- .32
47. Valladolid	- .12	.15	.26	- .21	.49	.27
48. Vizcaya	0.00	.48	.10	.75	.10	.35
49. Zamora	.53	- .02	.32	0.00	.52	- .06
50. Zaragoza	- .30	.08	0.00	.02	.23	.22

(a) = agriculture
(b) = all sectors

Sources: Annuario de Estadistica Agraria, 1965, 1977; Banco de Bilbao, 1975; OECD, 1970, 1975

which makes a new synthesis of surplus transfer data in the agricultural sector will be presented at the end of the chapter.

The data on gross income and gross value added (Tables 9.1, 9.2, 9.3) for each region or province both in terms of the agricultural sector and for all sectors, clearly show the pattern of regional differentiation in the three countries. The figures in columns (a) represent differences between the share of the region's GAI in total gross agricultural income and the share of the region's

Map 9.3: SPAIN: The Geographical Distribution of Receiving, Donor and Balanced Regions in the Agricultural Sector (GAI - AGVA)

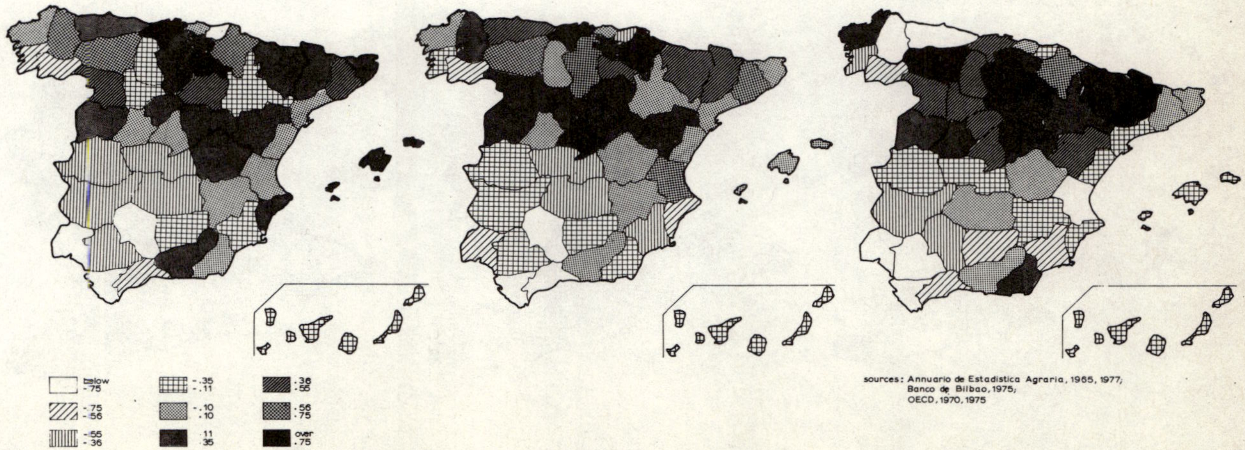

sources: Annuario de Estadistica Agraria, 1965, 1977;
Banco de Bilbao, 1975;
OECD, 1970, 1975

Map 9.4: SPAIN: The Geographical Distribution of Receiving, Donor and Balanced Regions Across All Sectors

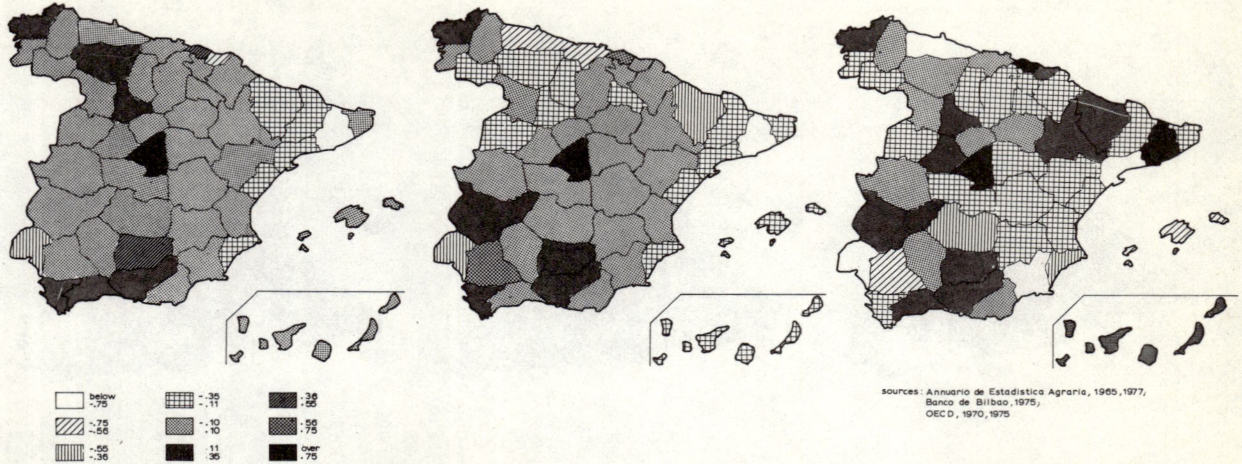

sources: Annuario de Estadistica Agraria, 1965, 1977, Banco de Bilbao, 1975, OECD, 1970, 1975

Table 9.3: Provincial Differences between Gross Income and Gross Value Added in Greece

	Percentage (Gross Income) - Percentage (Gross Value Added)		
	1955	1965	1975
	(a)	(a)	(a)
1. Athens Region	.75	1.30	1.60
2. Aetol-Akarnania	.11	.27	.30
3. Boetia	0.00	.18	.65
4. Euboea	- .10	0.00	.10
5. Evritania	- .25	- .10	0.00
6. Phthiotis	.10	.15	.13
7. Phocis	0.00	0.00	.12
8. Argolis	.35	.41	.90
9. Ahaia	.70	.90	.85
10. Arkadia	- .10	0.00	- .35
11. Ilia	.25	.12	0.00
12. Corinthia	.30	.36	.35
13. Laconia	- .15	0.00	- .75
14. Messenia	.25	.10	.12
15. Ionian Islands	0.00	- .11	- .33
16. Arta	.12	0.00	0.00
17. Thesprotia	0.00	- .33	- .65
18. Yannina	.12	.10	0.00
19. Preveza	0.00	- .10	- .15
20. Karditza	.18	.12	0.00
21. Larisa	.35	.46	.65
22. Magnesia	.21	.31	.61
23. Trikala	0.00	.10	- .12
24. Grevena	0.00	- .10	- .12
25. Drama	- .12	- .10	- .15
26. Imathia	.11	0.00	- .10
27. Thessaloniki	.45	.66	1.05
28. Kavala	.12	- .22	- .30
29. Kastoria	- .31	- .10	- .31
30. Kilkis	0.00	- .29	- .45
31. Kozani	- .10	- .65	-1.12
32. Pella	0.00	0.00	- .12
33. Pieria	0.00	- .36	- .33
34. Serres	.35	.76	0.00
35. Florina	- .71	.61	0.00
36. Khalkidiki	- .15	- .12	.21

Table 9.3: Provincial Differences between Gross Income and Gross Value Added in Greece (continued)

	Percentage (Gross Income) – Percentage (Gross Value Added)		
	1955	1965	1975
	(a)	(a)	(a)
37. Evros	0.00	- .30	- .28
38. Xanthi	- .76	- .35	- .31
39. Rodopi	- .60	- .75	- .95
40. Aegean Islands	.12	- .65	-1.20
41. Iraklion	.25	.39	.85
42. Lasithi	0.00	.11	.21
43. Rethymnon	- .10	0.00	.12
44. Canea	.21	.11	.10

(a) = agriculture

Sources: NSSG, 1951, 1961, 1971; Agricultural Statistics of Greece, 1963, 1969, 1975; Agricultural Bank of Greece, 1976; OECD, 1971, 1979; TEE, 1975.

AGVA in total gross national value added. Similarly, figures in columns (b) show differences between the region's share in total national income and the region's total contribution to gross national product. The figures in columns (a) and (b) can be viewed - given the data limitations - as an approximation of a GTV index, first within the agricultural sector and the second for the region or province as a whole, i.e. for all sectors. For example, in 1955 the province of Leon in northern Spain had 0.62 for its agricultural sector, which means that it received 0.62 per cent more in terms of the total gross agricultural income than its actual contribution in the total national gross agricultural product. In the same year, the entire provincial economy of Leon showed 0.14, which means that all its sectors received 0.14 per cent more income than their actual productive contribution in

Map 9.5: GREECE: The Geographical Distribution of Receiving, Donor and Balanced Regions in the Agricultural Sector (GAI - AGVA)

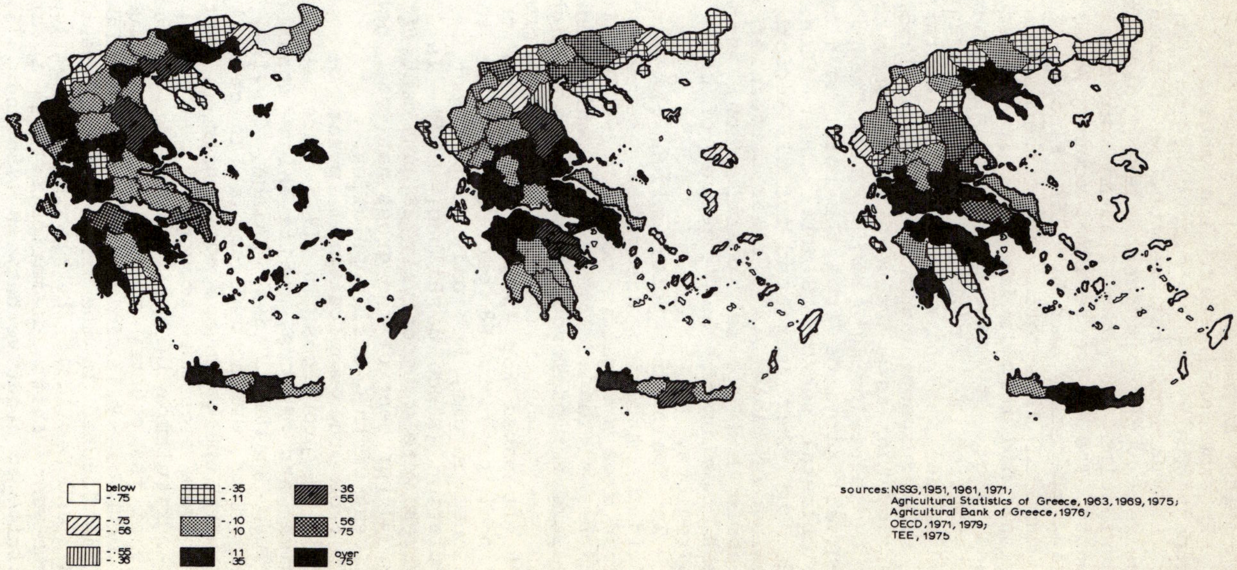

sources: NSSG, 1951, 1961, 1971;
Agricultural Statistics of Greece, 1963, 1969, 1975;
Agricultural Bank of Greece, 1976;
OECD, 1971, 1979;
TEE, 1975

the total gross national product. Leon was clearly a 'receiving' region.

In the same year, the region of Puglia in southern Italy was a 'donor' region. It had -0.85 for its agricultural sector, which means that it received 0.85 per cent less than its actual contribution in the Italian gross agricultural product. The entire regional economy was also losing during the same year by -15%. Finally, in 1955 the province of Thesprotia in northern Greece (Epirus region) had zero in its agricultural sector, which means that it managed a balance between its contribution to the total Greek gross agricultural product and its share in gross agricultural income. In this respect and for the year 1955, one can classify Leon as a 'value-gaining' province, Puglia as a 'value-losing' region and Thesprotia as a 'neutral' province from the point of view of their agricultural sector. Similar observations could be made for other years and for all sectors combined, showing in Table 9.4 a clear pattern of increasing economic and spatial polarization, with certain areas growing at the 'expense' of other provinces and regions.

In Spain the polarization of Barcelona and Madrid is striking across all sectors. Barcelona jumped from a negative position (-0.76) in 1955 to a high position (3.15) in 1975, while Madrid during the same period moved from 1.15 to as high as 6.20. When we include Vizcaya, these three provinces in 1975 accounted for 45 per cent of the total gross national product, 68 per cent of the total gross national income, but only 25 per cent of the total population (Banco de Bilbao, 1975). In 1955, these figures were respectively 38 per cent of the total GNP, 51 per cent of the gross national income and 16 per cent of the total population.

From the agricultural point of view, high receiving provinces in 1975 were Leon, Navarra, Lerida and Burgos, while high donor provinces include Huelva, Cadiz, Malaga, Sevilla, Valencia, Jaen, Lugo, Oviedo and Murcia. In general, most of southern Spain is a donor region, a pattern which has become even more pronounced over time. The index for agriculture in the 17 provinces in the southern half of the country moved from a net value of -6.41 in 1955 to -9.15 in 1975, with the highest donor values focused in the extreme south-west (Huelva, Cadiz and Sevilla). The provinces of Galicia and the north-west shift dramatically as a group from positive to negative between 1955 and 1975, led most dramatically by Lugo and Oviedo. In contrast, a pronounced regional concentration of high receiving regions, led by Leon and Burgos, develops over time in the north-central area. The highest ranking regions for all sectors (Madrid, Barcelona and Vizcaya) tend also to have positive values in agriculture, but except for Madrid in 1965 these are not particularly high.

In Greece, the figures verify the dominant position of the Athens region (1.60 in 1975) and Thessaloniki (1.05). These two provinces account for 41 per cent of the total population, 51 per cent of the total gross national product and 69 per cent of the total gross national income.[6] Other high receiving provinces in agriculture in 1975 are Argolis, Ahaia, Iraklion, Larisa, Beotia and Magnesia; while high donor provinces include the Aegean Islands, Kozani, Rodopi, Thesprotia and Kilkis. The dramatic decline of the Aegean Islands is particularly noteworthy.

In Greece, the spatial pattern and the location of donor and receiving provinces is somewhat different compared with Spain, (see Map 9.5). Taking into account the smaller size of the country, there is no clear distinction here between north and south, but there exists an 'S'-type corridor of receiving areas connecting Ahaia province in the Pelopennese, via the Corinth Canal to Athens and then to Magnesia, Larisa (in Thessaly) and Thessaloniki in the north. In 1955, this corridor development pattern was not yet clear. From 1965 onwards, however, a strong spatial differentiation appeared - both in terms of location and net figures - with high receiving provinces being Ahaia, Athens and Thessaloniki in 1965 and extending to include Argolis, Beotia, Larisa, Magnesia and Iraklion (in Crete) in 1975.[7]

In Italy, regional differences are less pronounced. The distinction between north-centre-south, however, is still noticeable. The three highest ranking, value-gaining regions in terms of total income are located in the north and centre (Lombardia, Piemonte and Emilia-Romagna), with fourth-ranked Lazio in the centre. The same four regions also rank highest in agriculture in 1975. More so than in either Greece of Spain, the leading value-gaining regions in Italy have remained highly stable over time.

The south generally contains the major value-losing regions, but over time there have been some noteworthy shifts. While such central regions as

Toscana, Umbria and Marche move from value-gaining to value-losing (in agriculture) between 1955 and 1975, several regions in the south (e.g. Puglia and Basilicata) change in the opposite direction. Sicily remains a value-losing region in agriculture in 1975, but the magnitude of this 'drainage' is significantly reduced. More dramatically, Sicily jumps from -0.80 across all sectors in 1955 to +0.10 in 1975. The other major reversal in the overall regional economy occurs in Veneto (+0.80 to -0.15), due most likely to the industrial mix found in the region (see, Indovina, 1977). One can broadly see in Italy, therefore, a stable core of accumulation in the central regions of the north, the beginnings of significant decline in most of the centre, and indications of improvement in a few regions of the south.

There is some evidence that the net differences between GAI and AGVA in the various regions and provinces in Spain and Italy are slowly being reduced. In other words, there is a tendency towards equalization of the conditions of production and exchange in the agricultural sector in the regions and provinces of the two countries. The opposite is true, however, when we take net differences for all sectors, where figures show an increasing tendency towards differentiation. For the Greek case, net agricultural regional figures show a tendency towards increasing differentiation. Evidence of these tendencies is shown in Table 9.4.[8]

Table 9.4: Coefficients of Variations of Net Figures in Agriculture and in All Sectors

		1955	1964	1975
Italy	(a)	.66	.64	.58
	(b)	.59	.63	.65
Spain	(a)	.71	.65	.69
	(b)	.51	.55	.72
Greece	(a)	.45	.49	.75

(a) = agriculture
(b) = all sectors

Sources: Tables 9.1, 9.2, 9.3

9.2 ALTERNATIVE REGIONALIZATION

The discussion of the GAI-AGVA index has shown the relationship between uneven development and the agricultural sector. Further investigation of similar trends in other sectors and the complex interaction between sectors is essential to complete the picture of uneven regional development in Italy, Spain and Greece. Having examined the agricultural sector in detail, however, I will propose an alternative regionalization for southern Europe based on agriculture. Again this is not a substitution for other possible regionalizations, but primarily an illustration of suggested methodology and a summary synthesis of previous sections.

The rationale for any particular form of geographical summary should be related to its usefulness in analysing the political effects of the patterns described (Massey, 1978a). In this respect, the regionalization presented below serves primarily a <u>political</u> goal; it shows exploitative regional relations existing in the agricultural sector which are not always clearly visible in basic statistical summaries. At the same time, it points toward an important arena for political action and struggle which, under a progressive leadership could help raise territorial consciousness among the population of both value-gaining and value-losing regions. Thus, 'region' may mean many things - in this case both a coherent spatial entity in terms of social, political and cultural/ideological relations and a geographical aggregate based on the GAI-AGVA index. Such an approach avoids a priori assumptions concerning the position of individual regions within the social and spatial division of labour of a particular social formation. The identification of a group of regions as 'core' and another as 'periphery' will come about as an effect of analysis, rather than as a point of departure. The combination of these effects produces a complex of spatial variation which is the empirical phenomenon with which regional analysis and regional policies is faced (Lipietz, 1977).

Using the data on regional differentiation in terms of the agricultural sector, I have classified the various provinces and regions of the three countries into four zones, using the following categorization:

<u>ZONE 1: Core regions</u>: Provinces having 0.56 and

over in their GAI-AGVA index.

<u>ZONE 2: Upward semi-peripheral regions</u>: Provinces having 0.11 to 0.55 in their GAI-AGVA index.

<u>ZONE 3: Downward semi-peripheral regions</u>: Provinces having -0.35 to 0.10 in their GAI-AGVA index.

<u>ZONE 4: Peripheral regions</u>: Provinces having below -0.36 in their GAI-AGVA index.

In Italy (Map 9.6), the pattern of regionalization shows that between 1955 and 1975 Emilia-Romagna and Piemonte retained their high-ranking receiving position, forming a stable core in the north, but the remaining Italian regions show a considerable degree of upward and downward movement: 12 out of 20 regions shift at least two zones. For example, Lombardia jumped from Zone 3 to Zone 1, and Sicily from Zone 4 to Zone 2. In contrast, 'traditional' commercialized agricultural regions, such as Veneto (north), Toscana and Marche (centre) were not able to compete with cheaper southern labour costs and dropped in 1975 from high to low ranking positions (Marche as low as Zone 4).

In Spain (Map 9.7) the regionalization follows a different pattern from that in Italy. Upward and downward regional shifts seem to have mainly benefited northern provinces. The majority of high-ranking receiving provinces in agriculture are located in the north, with the exception of Madrid, Granada and Las Palmas. In addition, the majority of the high-ranking donor regions are located in southern Spain, with the exception of Pontevedra, Valencia, Lugo and Oviedo (which verifies changes in the development gap between north and south) as in the case of Italy. During the period under study, data show a domination of northern agriculture over the rest of Spain, a conclusion which has also been reached from another point of view by OECD (1971). An extreme case of upward shift was Madrid (from Zone 3 to Zone 1); while for downward shifts, Lugo dropped from Zone 1 to Zone 4, Barcelona dropped from Zone 1 to Zone 3, and Tarragona from Zone 1 to Zone 3.

In Greece (Map 9.8) the regionalization pattern is quite stable for high-ranking receiving provinces forming the core, and very unstable for the rest of the country. The first eight provinces, with the exception of Iraklion and Serres, remain the same in 1955 and 1975.

A first observation, common for the three

Map 9.6: ITALY: Alternative Regionalization

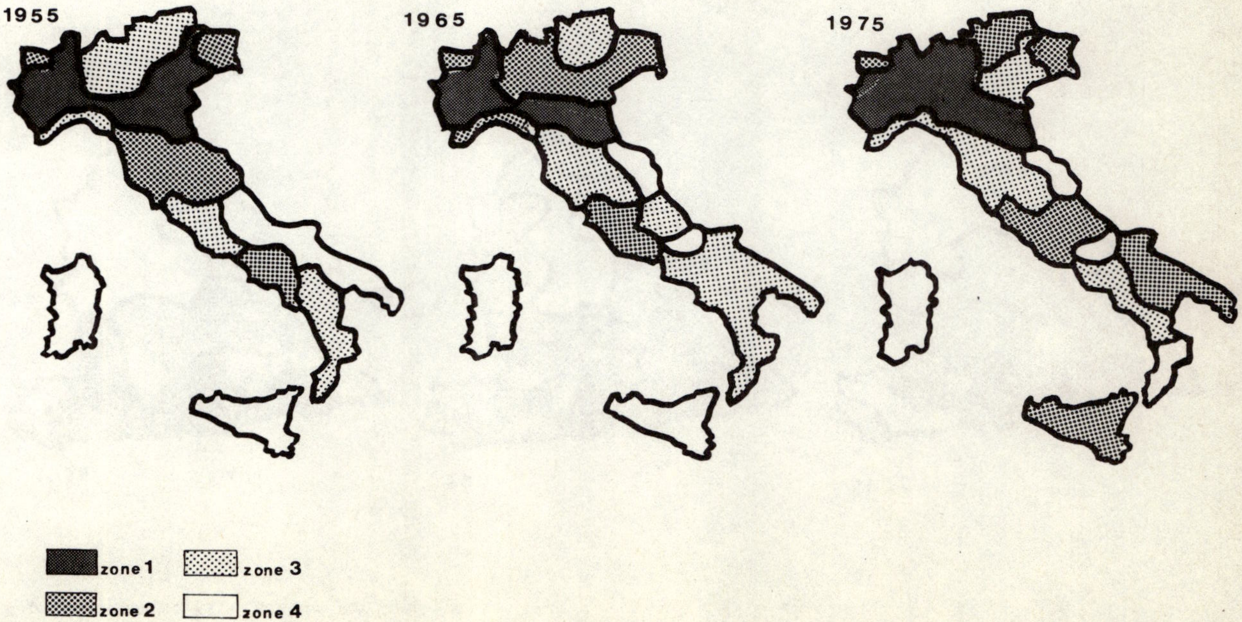

Map 9.7: SPAIN: Alternative Regionalization

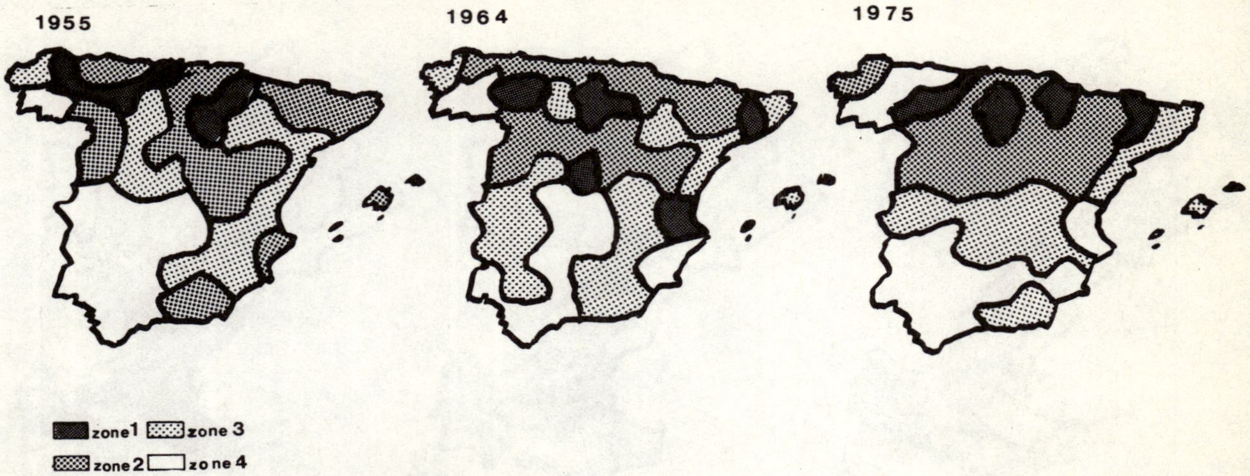

Map 9.8: GREECE: Alternative Regionalization

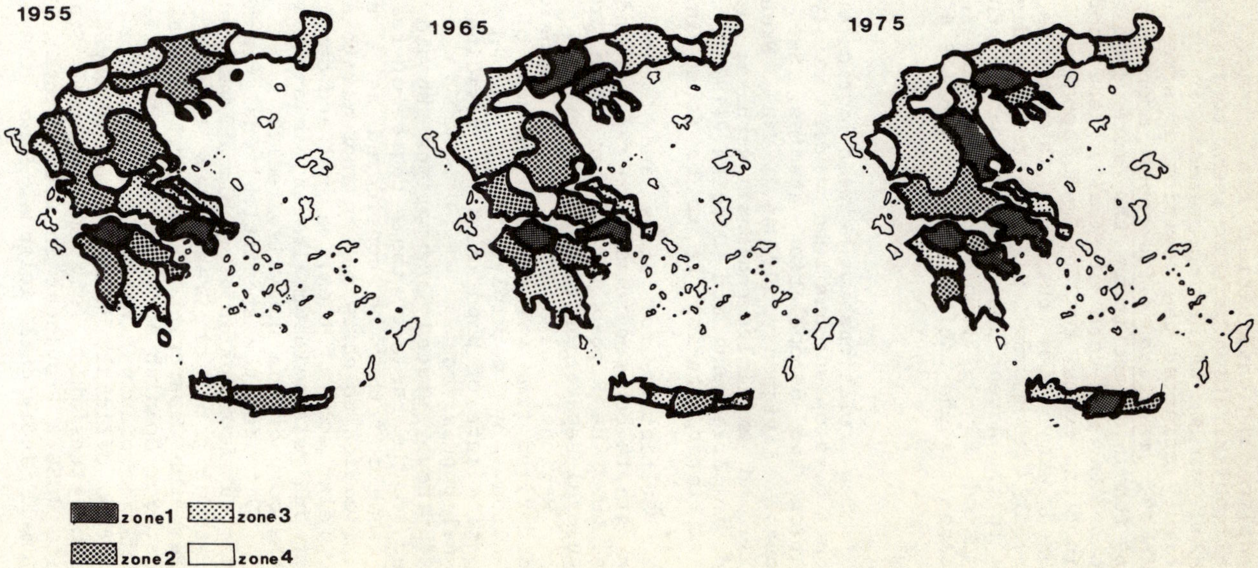

countries, could be that from 1955 to 1975 their respective regionalization has changed but the overall polarization pattern seems to remain stable. In other words, while improvements in productivity positively affected the position of some individual provinces, relative differences among the four zones remain, benefiting the more industrialized regions. In the case of Italy, the north-south distinction seems to be less apparent in 1975 compared with 1955, but this is not the case in Spain. In Greece, the 'S'-type corridor development close to the Aegean Sea remained constant throughout the period 1955-1975, while regional differentiation increased, as the coefficients of Table 9.4 show.

NOTES

[1] The relationship between agricultural and regional development has been studied mainly through migration and depopulation studies. See, Kayser, Péchoux, Sivignon (1971); Van Nieuwenhuije (1972). An explicit regional emphasis is to be found in Friedmann and Douglas (1975); while a Marxist interpretation is to be found in Lipietz (1977).

[2] Statistics for Italy and Greece did not differentiate between returns to capital and labour, while Banco de Bilbao (1977) for Spain provided a very useful separation between the two.

[3] Regional value added can be considered as being derived from value added per head of occupied population, by multiplying the latter by the occupation rate of population and dividing by total regional population.

[4] These stated assumptions should not be judged on the basis of their degree of realism but they should rather be considered as a necessary simplification to carry out an approximate analysis. If we relax some of them, we just add complications to our analysis without altering in any significant way the core of the analysis itself.

[5] Data for the statistical analysis have been calculated from official publications. Additional statistical data for Greece have been provided by the Research Department of the Agricultural Bank of Greece. For Spanish and Italian geographical names, I kept the original native spelling. For Greece I used the official (though not always correct) English translation of the National Statistical Service (NSS). Provinces in Greece total 51, but 9 are small individual islands. Since they did not

have important internal differences, I decided to put them together in two groups, the Ionian and Aegean Islands, thus having 44 provinces in total. I am also aware of the disaggregation problem in Italy: fewer regions compared with provinces in Spain and Greece. But the number of provinces in Italy is very high - 91 - and tends almost to coincide with the number of 'communities' in Spain and Greece.

[6] Athens leads in all time periods. Thessaloniki is third in 1955, fourth in 1965 and second in 1975.

[7] Iraklion is connected with a ferry boat line with the transportation axis via Piraeus port.

[8] Coefficients of variation were calculated by the formula:

$$V = \frac{\dfrac{\sqrt{\Sigma_i (X_i - \bar{X})^2}}{N}}{\bar{X}}$$

where X_i = value of each variable for the region, $\bar{X}$ = the mean, N = number of regions. See Blalock (1972).

Chapter 10
PRODUCTION, LOCAL SOCIAL STRUCTURES AND POLITICAL/ IDEOLOGICAL MEDIATION

The statistical and geographical analysis of the differences between GAI and AGVA in Italy, Spain and Greece has illustrated the dynamics of uneven regional development through the agricultural sector. It has not shown, however, why certain regions and provinces are 'value-gaining', 'value-losing' or 'neutral' and how they change their position over time.

10.1 SOCIAL CHARACTERISTICS OF AGRICULTURAL PRODUCTION AND STATE ECONOMIC INTERVENTION

Looking first at the farm structure, two basic types of farm enterprise are present, although unevenly distributed, in the agricultural regions of southern Europe (see also, Camilleri, *et al*., 1977; Louloudis, 1984; Mottura, 1980):

a. the capitalist farm, typically specialized in cereal and industrial crops is found mainly in well-irrigated lowlands and is largely based on the employment of wage labour. Its surface area is usually greater than that of the average farm in the respective region. These farms form a considerable proportion of agricultural enterprises in Spain and Italy, while their presence in Greece is less important.
b. the peasant farm specialized in Mediterranean products and, although present in the lowlands, is more clearly predominant in high and mountainous zones as well as on the islands. This form of agricultural production is dependent on family labour, it is widely spread all over southern Europe and, as already noted, it is the focal point of the ongoing debate about the agrarian question in the three

countries.

Going beyond this general description, a more elaborated typology of farm enterprises is given in Figure 10.1. The agrarian structure, composed in each geographical area, is formed by one or more of the five main farm types with their corresponding characteristics. While the typology is presented as a fixed set of forms of production, labour market, social classes, etc., it is not meant to offer a static picture of the agrarian structure in the three countries. On the contrary, it permits the construction and investigation of the changing mix of forms of agricultural production in the various Mediterranean regions, which in turn influences differentially localized accumulation patterns.

The capitalist farm is subdivided into _traditional_, _backward_ and _modern/dynamic_ capitalist farming (Moreno and Rodriguez, 1979). The first (traditional) is identified with a corresponding class structure composed of a dominant but declining landed bourgeoisie and dominated farm-workers. With a minor presence in Greece, these farm enterprises are normally of considerable size (in southern Italy and southern Spain, over 200 hectares) and located in well-watered lowlands. They use both regular and seasonal labour with close to national averages of land/labour ratios. With adequate, though sometimes old mechanization, they tend to specialize in products with which economies of scale can be obtained. They are located mainly in central and southern Spain, southern Italy and central Greece.

Backward capitalist farms are usually medium or relatively small in size and are found in all three countries. Present in both lowlands and mountainous areas, they seem to constitute a transitional stage either towards absorption by the most dynamic farms or towards progressive abandonment, especially in cases of regional depopulation, ageing, inadequate rural infrastructure, etc. These farms are less mechanized and are oriented to 'traditional' crops, such as olive oil, grapes and tobacco. In all three countries these types of farms are not concentrated in specific regions, with the partial exception of Thessaly, in Greece.[1]

Finally, _modern/dynamic_ capitalist farms are mainly of medium size and include advanced and capital-intensive structures, specializing in short-cycle crops (usually in greenhouses), fruit and vegetables, flowers, cattle and other animal quick-fattening processes. They may operate on rented

Figure 10.1: Typology of Farm Enterprises in Southern Europe

	Types of Farms	Local Labour Market	Local Class Structure	Capital Composition and General conditions of production	Forms of state's Dominance
	Traditional Farming	Wage Labour Latent proletariat likely to be unionized	Declining landed bourgeoisie Farm-workers	High land/labour ratio Relative mechanization Good irrigation Adequate infrastructure	Price and wage regulation Terms of trade Marketing
The Capitalist Farm	Backward Farming	Low-wage labour Share-cropping Under-unionized	Rural-urban 'old' petty-bourgeoisie Absentee management Farm-workers	High land/labour ratio Low mechanization Little irrigation and weak infrastructure	Price and wage regulation Taxation policy
	Modern Dynamic Farming	Part-time and seasonal workers Participation in the informal economy - Usually excluded from solidaristic organizations	Corporate capital Cooperatives and municipal ownership Part-time workers	Low land/labour ratio High mechanization Developed infrastructure and good irrigation Proximity to big urban centres	Price regulation Marketing Acceptance of informal economy Incentives to capital

The Peasant Farm	Subsistence/ traditional Farming	Family labour Part-time work increasing feminization and ageing	Poor Peasants Semi-ptoletarians	High land/labour ratio Very low mechanization Good irrigation Undeveloped infra-structure	Price regulation Direct subsidies Migration policies
	Dynamic/ specialized Farming	Family labour Independent small-business-subcontractors Participation in the informal economy. Non-solidaristic attitudes	Rural-urban 'new' Petty-bourgeoisie Cooperatives Part-time workers	Medium land/labour ratio Adequate mechanization Good irrigation Adequate infra-structure Dense settlement network	Contract farming Banking loans Special planning programmes

land - contrary to the other two forms - and are highly adaptable to market demand and fluctuation. They have the better access to technology, finance and markets, while they tend to use sporadic, marginal short-term employment. These farms seem to be expanding in northern and central Italy, in parts of Andalusia, northern and eastern Spain, while in Greece there are indications of such development in the Peloponnese, Crete and central Macedonia, due to interests from shipowner capital and cooperatives financed by the state.[2]

In terms of the peasant farm, there can be distinguished two principal subdivisions: <u>subsistence/traditional</u> and <u>dynamic/specialized</u> peasant farming. The first form is overwhelmingly determined by the contradiction between the availablity of land and the availability of family labour (Pugliese, 1979). These farms are small or very small in size (below 5 hectares) and very fragmented in terms of parcels for cultivation. They are present in high and mountainous regions operated by ageing family members (predominantly female labour) and tend to be highly resistant to structural changes. In terms of class structure they are mainly constituted by poor peasants and/or semi-proletarians depending on the central state for price regulation and loans and on local authorities for welfare services (Camilleri <u>et al</u>., 1977; Vergopoulos, 1975).

In contrast, dynamic/specialized peasant farms, while they are dependent on family labour, state price regulation and loans from the banks, combine capital-intensive farming in small size farms, through a specialization on certain short-cycle 'shoft' crops like vegetables in greenhouses, flowers and fruit. In many cases they perform their cultivation programmes under a specific contract either with a state agency or an agricultural processing industry. With expanding EEC markets, they seem to be growing fast in such well-watered regions as southern Peloponnese, central Macedonia, Crete, central-southern Italy, Andalusia, Almeria and Catalonia. In terms of class structure this type of farm consists mainly of relatively well-to-do peasants with one or more family members having a regular job outside the agricultural sector. Use of seasonal non-family labour is also not uncommon.

This farm typology is a first indication of <u>social differentiation</u> in the countryside, especially when it is further related with two important issues: <u>regional labour markets</u> and <u>local general</u>

conditions of production and reproduction. This double relation directly affects incomes derived from control of the means of production, including personal labour (production income), income from the sale of labour power (wage income) and income derived from welfare services (welfare income).

Starting first with local labour markets, regional differences are important in terms of both supply and demand of agricultural labour due to the uneven decline of agricultural population.[3] However, as many studies have noticed (OECD, 1973, 1979), more important than the absolute decline in agricultural population was the decline in the number of work days actually performed by each worker in agriculture. This is particularly manifested in the growth of second jobs in rural areas and part-time farming. In Italy during the period 1966-1975, about 60 per cent of peasant farms absorbed less than one person's annual labour per annum and a little under one-third of farms generate a demand of less than 50 days of labour per annum (INEA, 1975). In Greece during 1961-1971 72 per cent absorbed less than one person's annual labour per annum, while for Spain the same figure was 62 per cent.

If we take into account the nature of the average family unit in peasant farms, then these figures give us an indication of underemployment in the agricultural sector. For the period under study, underemployment reached around 50 per cent in Italy (50 per cent in Abruzzi and Molise, 80-90 per cent in Calabria and Apulia), over 65 per cent in Greece (70-80 per cent in central Greece, 90 per cent in Epirus), and around 56 per cent in Spain (90 per cent in Galicia, 65 per cent in Andalusia). Underemployment coincides, however, with seasonal demand for both qualified and non-qualified agricultural labour. This situation is particularly aggravated in Greece, where during the harvest of specific products (e.g. olives, fruits, grapes) labour is imported from other regions (e.g. from Macedonia, to the Peloponnese and Crete) or from abroad, while local farm-workers prefer to work in non-agricultural activities (e.g. in tourism).

For those people working part-time in agriculture and looking for another job, secondary employment is highly heterogeneous and spatially differentiated. In lowlands where intensive crop production is taking place seasonal labour in capitalist farms is possible (e.g. central-southern Spain, central-southern Italy, central Madeconia).

In the majority of other cases, however, farm-workers and peasant landholders usually seek part-time employment in non-agricultural activities. Depending on regional productive structures, physical characteristics (e.g. proximity to the sea) and job qualifications, these activities include the building industry, public works, tourism, seasonal services to the state sector and industry. In all three countries, part-time work outside agriculture also involves important seasonal and periodic migration out from the home region. Often this represents a first move towards permanent migration (Kayser, 1970). Earnings from non-agricultural work are much less frequently reinvested in farming; rather they are used for direct consumption, savings or opening an individual small enterprise. Finally, important changes took place in terms of regional labour markets due to a new pattern of diffused industrialization in rural areas. In many parts of Spain, Italy and Greece (as well as in Portugal), the crisis in the early 1970s generated a phenomenon known as <u>endogenous industrialization</u> (see also, Granados, 1984; Hadjimichalis, 1984).

This process refers to non-agricultural activities developed without the direct intervention of the state (as in the case of the Cassa per il Mezzogiorno) or the decentralization of large enterprises (as in the case of FIAT). Endogenous industrialization is based on the use of the areas' own resources and bears a very clear relation to the phenomena of small enterprises and informal economy. Informal economic transactions include 'black work' (that is a job without legal coverage), part-time work (less than 17 hours a week), home by-piece production and buying inputs and selling products and services without official registration, that is avoiding taxes (see also, Vinay, 1985). These developments have increased agricultural incomes while at the same time have kept wages low in industries as the available family income has multiple sources, and unionization and solidaristic attitudes are kept away from small enterprises. This has favoured technological changes in agriculture and had a direct impact on the kinds of agricultural crops grown in certain regions such as Beotia, Corinth and eastern Macedonia in Greece, Calabria, Molize and Liguria in Italy and in Valencia, Alicante and Lugo in Spain.

The second important issue in terms of social differentiation in the countryside is <u>local general</u>

conditions of social production and reproduction. These conditions include those services and infrastructure provided by the state and local authorities, which mainly tend to increase the productivity of labour, to reduce the cost of reproduction of labour power and to shorten the turnover time of capital. As in cases of productive structures and labour markets, the historical evolution of regional differentiation plays an important role. In these conditions, however, the role of state and local authorities acquires greater importance. Such intervention in rural areas is concentrated mainly in four major areas:

a. In the rural technical infrastructure such as irrigation projects, land improvements and consolidation, crop storage, and rural roads. These programmes tend to be organized to promote the logic of comparative advantage, thus favouring regions where some cumulative 'success' is already present. Other technical and social subsidies (e.g. rural health service in Greece) which do not follow market criteria, tend to benefit other sectors more than agriculture. One reason for this is the demographic situation in backward rural regions, where old peasants cannot or do not want to take advantage of these subsidies.
b. In the national rural settlement policies (villages, small and medium size towns) where the majority of social and commercial services are concentrated. These services - as it is widely documented - tend to follow a certain hierarchy which usually means that the lower ranking villages are underserved (e.g. travelling long distances to reach a doctor, a bank or a school in a distant village or town) and have low living standards. Nevertheless, state and local activities' efforts succeeded by the late 1970s in developing a rather sufficient scheme of rural communications and settlement structure which favoured modernization in agriculture and endogenous industrialization in certain areas. The latter facilitates the articulation of the system of small firms in rural areas with small urban centres.
c. In the marketing of agricultural commodities and the supply of material inputs which remain very inefficiently organized. One of the major weaknesses of southern European agriculture is

the lack of local supportive institutions for farming which have greater economies of scale than does the direct process of production. These supportive institutions include the wholesale structure, transportation and distribution facilities, financing, research and development, continuing farming education, etc. The paucity of these institutions (usually controlled by the state and para-state organizations) means for southern Europe dynamic farmers and inadequate infrastructure for the reduction of the turnover time for their capital.

d. Finally, in almost all southern European countries an important gap exists in means of collective consumption between urban and rural areas, including housing, health facilities, educational opportunities, employment diversification, recreation and the position of women. In Greece (1975) for example, while urban areas have 2.6 doctors per 1,000 people, rural areas have only 0.35. The average figure for doctors per 1,000 people in the EEC is 3.8 for urban areas and 2.7 for rural areas. In Italy (1975) housing conditions in the north, with 1.8 persons per room, are still better compared to the south, with 2.9 persons per room. In Spain and Greece rural areas have the highest percentage of illiterate persons. In 1971, as much as 14.2 per cent of the total population of Greece was illiterate (222,000 men and 817,300 women) and 72 per cent of them lived in rural areas (Eliou, 1978). Similarly in 1974, 10.1 per cent of the total population of Spain was illiterate, with 61 per cent of them living in the countryside (Estatistica Agraria, 1978).

Local general conditions of social production and reproduction which are shaped by the contradictory policies of the state and local authorities tend to be highly spatially heterogeneous. In fact these conditions provide an alternative framework for an analysis of uneven regional development moving beyond productive structures alone and seeing production and consumption as a dialectical unity (Bleitrach and Chenu, 1979). It is these variable conditions which largely determine the location of firms which are seeking new sites for exploitation. In relation to our farm typology, a broad correlation seems to exist between traditional/peasant and backward capitalist farms and low or inadequate development of local general conditions. The majority of regions where these farm types are dominant (with some notable exceptions, as the Peloponnese in Greece and Apulia and Molise in Italy) is marked by inefficient irrigation, underdeveloped transportation and settlement systems and low living standards.

Thus a vicious cycle seems to exist, the exit from which requires other forms of policy and totally different kinds of practices from those traditionally applied by state agencies.

10.2 CHANNELS FOR THE GEOGRAPHICAL TRANSFER OF VALUE IN SOUTHERN EUROPEAN AGRICULTURE

Various studies of the agrarian question in southern Europe and elsewhere argue that surplus produced by the small peasant producer is transferred out from agriculture and that part of this surplus is appropriated by industrial capital (see Amin and Vergopoulos, 1974; Bartra, 1975; Lipietz, 1977; Caballero, 1979). With the exception of Lipietz, however, all other authors approach the problem from a sectoral point of view, missing the important connection with space and, through this, with uneven regional development.

Although there are differences between the writers on this subject, most tend to deal with this transfer from the small peasant producers to industrial capital in terms of the relations of unequal exchange that take place between the peasant sector and the urban industrial and commercial sector. Bartra (1975, pp.80-2, 1975, pp.136-7) and Vergopoulos (1975) for example, argue that the transfer of surplus from the peasantry to the industrial bourgeoisie takes the form of an appropriation of the surplus labour originating in peasant production obtained as a result of the low market prices paid by the urban-industrial sector. That is to say, since the peasants must often sell their products for prices below their value, part of the labour expended in the production of these products is not remunerated and is thus appropriated by the industrial or commercial 'consumers' who buy these products at low prices.

A similar view is addressed by Caballero (1979) who, writing for Spain, argues that the unfavourable terms of exchange suffered by peasants are a consequence of the structural articulation between the peasant mode of production and the dominant capitalist mode of production. Similarly, Lipietz (1977) presents in his work a spatial division of French agriculture into three major zones, according to the type or stage of articulation of the peasant mode of production with the CMP. Lipietz correctly points out that the CMP always needs an extra-economic relationship in order to implant itself in pre-capitalist formations. Economic mechanisms, he argues, are not sufficient to break up the closed production system of subsistence based organizations.

Unequal exchange under certain conditions does take place in agriculture as well as in other sectors within a country, being thus an important factor for uneven regional development (Lipietz, 1977; Sayer, 1977; Liossatos, 1979). But it is not the only factor influencing surplus transfer, nor necessarily the most important. It is simply one factor among many which must be identified in the light of a detailed analysis of the unity of production and circulation. Thus, building upon these observations and upon the discussion in the previous section, three major channels of GTV in the agricultural sector in southern Europe can be identified: direct GTV through financial transfers, 'living labour' transfers and indirect GTV through commodity transfers.

Financial Transfers

The first channel is related with financial flows when the marketed surplus is accompanied by a net transfer of investible resources out from agricultural regions. A net transfer of surplus out of agriculture will occur via the free market when three conditions are satisfied (see also, Griffin, 1979). First, the agricultural sector in a particular region must sell part of its production to other sectors, i.e. there must be a marketable surplus. Secondly, the agricultural sector must consume less than its total income, i.e. there must be net savings by the sector. Finally, investment in agriculture must be less than savings by agriculture.

Financial flows include both voluntary and involuntary savings transfer. Rural savings are voluntarily invested in other (mainly urban) regions, presumably because the rate of return on capital is higher than in the countryside. Landowners, the state (including the church), capitalist farmers, 'middle men' and various financial institutions (e.g. banks, usurious capital) are the key figures in financial transfers. Through their control of the land and commercial contracts they are able to appropriate the agricultural surplus and to invest it in urbanized regions, mainly in the building sector (Sakelis, 1985). In Spain, many large estate owners are also industrialists or involved in real estate and the tourist business, thus transferring easily savings from agriculture to other sectors.

Another example of voluntary transfers is through profit repatriation (which occurs among both regions and countries), which is particularly applicable in cases where international capital or national industrial capital is invested in agriculture. These cases include shipowner capital investment in high technology agricultural production in Greece (southern Peloponnese, Crete), industrial capital invested in Spanish and Italian agriculture (Povaley, Alicante, Andalusia) and finance capital invested in agricultural activities in all three countries. In these cases capital-intensive production takes place using low-paid farm-workers, often illegal migrants from Libya, as in Sicily and the southern Peloponnese. These firms are vertically integrated, controlling production transportation and the distribution of their products to foreign markets, mainly the EEC. In cases of lack or limited import of foreign currency from their sales, we have an extreme case of financial transfer (direct GTV) towards the international market.

Involuntary savings are 'extracted' from agriculture by a variety of means. They are affected by unfavourable movements in the intersectoral terms of trade against agriculture, sustained and reinforced by middlemen and commercial capital, despite some efforts by the state to control relative prices of strategic farming inputs and agricultural products' price controls. In this respect, control of the price system, and hence the intersectoral terms of trade, is the absolutely strategic mechanism for achieving involuntary reallocations of surplus between different sectors, social classes and regions. The previous points can be illustrated by the tendency of the index of prices received by farmers in southern Europe to be lower than the index of prices paid for agricultural inputs. Table 10.1 gives such a picture, showing how small is the general profit margin in southern European agriculture (ratio of index A to index B)

Table 10.1: Index of prices received and paid by farmers in Italy, Spain and Greece

ITALY	1967	1969	1971	1974
1. General Index A for prices received by farmers (1966=100)	101.8	107.0	113.3	156.3
2. General Index B for prices paid by farmers (1966=100)	97.9	102.9	106.1	139.4
3. Index A / Index B	1.03	1.03	1.06	1.12
4. Industry A/B	3.12	4.65	5.12	8.10

Source: Bolletino Nevile di statistica no. 7, 1974; UN Statistical Yearbook, 1975

SPAIN	1955	1960	1965	1970	1976
1. General Index A for prices received by farmers (1964=100)	53	82	116.5	128.5	237.8
2. General Index B for prices paid by farmers (1964=100)	74.1	84.3	103.3	114.6	207.8
3. Index A / Index B	.71	.97	1.12	1.12	1.14
4. Industry A/B	3.60	3.95	4.58	5.80	6.70

Source: Annuario de Estadistica Agraria, 1977; Annuario de Estadistica, 1978

GREECE	1968	1971	1973	1975
1. General Index A for prices received by farmers (1966=100)	102.6	112.6	166.9	171.3
2. General Index B for prices paid by farmers (1966=100)	101.2	107.8	193.5	161.2

Table 10.1: Index of prices received and paid by farmers in Italy, Spain and Greece (continued)

GREECE	1968	1971	1973	1975
3. Index A / Index B	1.01	1.04	.86	1.18
4. Industry A / B	2.10	3.25	4.15	5.80

Source: National Statistical Service, 1978; Agricultural Bank, 1980; National Accounts, 1978

which sometimes falls below 1, as in Spain in 1955 and 1969 and in Greece in 1973. For comparative purposes and for the same years the ratio of index A and index B for industry for prices received and paid by industrialists is presented. It is always higher than agriculture and it is growing at a faster rate.

'Living' Labour Transfers

The second major channel of GTV is related with the value transfer in the form of 'living labour', that is people who have migrated from rural areas and whose cost of reproduction was paid in their areas of origin (Kavouriaris, 1974; Psichogios, 1984). In the cost of reproduction of the labour force are included: (a) the part of family income devoted to child rearing, (b) the non-paid labour time spent by family members - especially women - for child rearing, (c) social expenses paid by the family and/or the community for education, health care, entertainment, etc., and (d) the investment opportunities lost because of the income and time spent for children. It is evident that these costs are very difficult to estimate. A number of issues such as the age of entry in the productive activit the years of education, the social expenses provided by the state and the role of social mobilizations, makes the exact calculation very complicated. Sauvy (1970) however made an approximate estimation for western Europe that the cost of reproduction of a worker with 10 years of basic education was equivalent to 9 years of work. Using a similar methodology, Loucopoulos (1979) estimated the cost of reproduction of a worker in Greece with 6 years basic education to be 6 years of work (during the decade 1951-1961). If we take the mean daily payment of a qualified worker in 1960 to be 150 drs. (US$3.75 in 1970 prices) and use the actual numbers of out- and in-migration in the various Greek regions, we have a first approximation of the value transferred via migration patterns. Table 10.2 gives such a picture.

We can see from the table that during 1951-61 only Attica had a positive migration balance. Of the 177,700 people, 120,700 were between 20 and 54 years old, capable of work. Using the hypothesis of Loukopoulos and taking US$3.75 as the mean daily payment, we find that Attica gained through internal migration the equivalent of US$760,410,000 in 1970 prices. We do not know the amount of migrants' remittances sent to rural areas, but related data from international migration show a big difference between the value of labour power and remittances, and there is no reason to suppose a different pattern for internal migration.[4] Another possible way to 'give back' to value-losing regions is through public spending (e.g. public works, social services, social security, etc.). But during the period under study the state and local authorities have not used this way to compensate for negative effects of internal migration. From Table 10.2 we can see that per capita public spending in Attica is high compared to other regions which have high negative migration differences, such as Epirus or the Aegean Islands. It seems therefore that Attica region was a net value-gaining region from the migration point of view, something which also corresponds with figures of GAI-AGVA (see Table 9.3).

When migrants change their sectoral working position, that is from agriculture to industry or services, we have an intersectoral transfer; when they continue to work in agriculture, but change geographical areas or farm structures, we have an intrasectoral transfer. It seems that the majority of movements in southern Europe belongs to the intersectoral category. Finally, all transfers can be analysed as surplus labour transfers because the reproduction of the labour force in rural areas in the period under study has taken place under conditions of petty commodity production.

Commodity Transfers

The third major channel of surplus transfer takes the form of '<u>dead labour</u>', embodied in agricultural commodities. The agricultural subsector as a social subsystem in each country is related with the

Table 10.2: Migration patterns and per capita public spending in Greece, 1951-1961

	In Migration	Out Migration	Net Difference	Per capita public spending
Attica	218.2	40.5	+177.7	2,150.5
Sterea Hellas and Euboea	31.7	51.9	- 20.2	3,610.5
Peloponnesos	15.5	75.5	- 59.1	2,310.6
Ionian Islands	2.5	16.5	- 14.0	640.3
Epirus	4.6	24.4	- 19.8	710.4
Thessaly	11.3	25.6	- 14.3	3,110.6
Macedonia	25.3	30.2	- 4.9	1,810.5
Thrace	6.0	6.6	- .6	630.2
Aegean Islands	6.2	33.1	- 27.0	412.5
Crete	5.1	22.8	- 17.7	615.5

Note: Population figures in thousands
Public spending in thousand drs.

Source: National Statistical Service, 1963; National Accounts, 1970

international economy and the EEC, with other national sectors and with various national and local institutions. More specifically it is related with (a) those who are selling inputs and services to agriculture including the banking sector, (b) those who are buying the agricultural products and (c) the state and local authorities. Furthermore, we have to make the distinction between production which takes place in large capitalist farms - where wage labour dominates - from production in small peasant farms where family labour is the rule. These two basic forms of production have qualitative different relationships with the 'outside world'.

In the first case, the capitalist farm, when a transfer occurs, this is mainly intersectoral and takes the form of surplus value transfer proper. There is also the case of intrasectoral transfer of value proper, when backward capitalist farming could lose value in favour of either other types of capitalist farms or towards dynamic/specialized peasant farms. This process represents a redistribution of surplus value between direct exploiters, that is between agrarian capitalists (intrasectoral) and/or between agrarian and industrial capitalists (intersectoral) (Goodman and Redclift, 1981). The case described here is one of many possible sets of circumstances in which relative price movements transfer surplus value between capitalists located in the same or different regions. I had described such transfers in Part I, Chapter 4, as part of the wider contradiction of equalization/differentiation. It is the case where regional differences in terms of general conditions of production play an important role. General conditions are acting as a source of 'quasi-rent' to be added as a coefficient to the cost price. Two important influencing factors here are unequal exchange and differential productivity.

Unequal exchange depends on differences in the organic composition of capital and in wage differentials. In the transformation of value into production prices and then into market prices - the case of capitalist farms - the gap between value and price is modified in favour of the farm enterprise with the higher organic composition of capital (intrasectoral transfer) or in favour of the sector with higher organic composition (intersectoral transfer). GTV through unequal exchange is applied paticularly when agriculture exchanges its products with other sectors, or when the state turns the terms of trade against agriculture by imposing price controls on agricultural products, and the use of multiple exchange rates that discriminate against agriculture.[5] These issues refer also to price and taxation policies controlled by the state. The previous points can be supported by data presented in Table 10.3 and Maps 10.3, 10.4, 10.5, 10.8, 10.9, 10.10, 10.13, 10.14 and 10.15. Table 10.3 shows the structure of wage differentials among agriculture, industry and services in the three countries, where agricultural wages remain always behind those of the other two sectors. Maps show the geographical distribution of wage differentials and regional differences in the organic composition of capital in agriculture, taking irrigation (irrigated land as a percentage of total cultivated land) and number of tractors per cultivated area as representative indicators.

In the case of differential productivity this major influencing factor of value transfer originates when a unit of product produced in a region will require more embodied labour than the national average of embodied labour consumed in the production of that commodity. Thus, the effective

Table 10.3: The structure of wages in Italy, Spain and Greece

Wages in Italy	1967	1969	1971	1974
Agriculture	65.1	68.1	68.6	71.2
Industry	98.2	101.2	106.8	118.5
Services	130.1	137.2	140.5	147.2
Total Italy (average of all sectors)	100.0	100.0	100.0	100.0
Wages in Spain	**1963**	**1971**	**1973**	**1976**
Agriculture	59.8	61.7	61.7	68.2
Industry	96.9	91.7	91.7	101.4
Services	127.5	123.5	121.2	131.2
Total Spain (average of all sectors)	100.0	100.0	100.0	100.0
Wages in Greece	**1967**	**1971**	**1973**	**1975**
Agriculture	58.1	59.6	60.3	67.3
Industry	81.2	85.2	90.2	98.3
Services	125.1	130.1	127.5	138.3
Total Greece (average of all sectors)	100.0	100.0	100.0	100.0

Sources: My own elaboration of data obtained from Annuario de Estatistica di Lavoro, 1978; Annuario de Estadistica, 1979; National Statistical Service of Greece, 1979

devaluation of embodied labour that originates in certain peripheral regions under conditions of lower productivity leads to value transfer to more productive ones. It also results in lower wages, which is a reflection of the normal function of the law of value in the formation of market prices. These points could be supported with data from Spain and Greece (see Tables 10.4, 10.5), where regional

Table 10.4: Agricultural differentiation by region in Spain (1975)

	Contribution to the national GAP %	Labour productivity national sectoral average = 100
Andalusia	20.9	110.7
Catalonia	8.1	133.0
Madrid	1.4	137.6
Valencia	8.9	119.8
Castilla-Leon	16.6	112.0
Galicia	10.2	50.2
Vasco-Navarra	5.4	158.1
Mancha	9.0	125.4
Canary Islands	2.7	90.0
Aragon	5.0	134.8
Asturias	2.7	62.2
Extremadura	4.7	86.3
Murcia	2.7	111.0
Balearic Islands	1.1	85.7

Source: Banco de Bilbao, 1975

Table 10.5: Agricultural regional differentiation in Greece (1973)

	Contribution of the region to national GAP %	Labour productivity national sectoral Average = 100
Athens Region	1.1	127.0
Macedonia	27.5	128.5
Peloponnese	18.2	85.2
Sterea Hellas	16.0	81.2
Thessaly	13.2	109.2
Epirus	4.2	53.6
Thrace	6.0	84.4
Crete	7.6	90.2
Aegean Islands	4.0	61.2
Ionian Islands	2.2	70.2

Source: Agricultural Bank of Greece, 1973

differences in labour productivity acquire important dimensions.

In the second case of surplus transfer through commodities - when production takes place in small peasant farms - a transfer occurs as a result of articulation between petty commodity production and the CMP. In this case, surplus transfer takes the form of surplus labour and can be both inter- and intrasectoral. The form of appropriation in petty commodity production can be conceptualized in terms of devalorized labour time (Goodman and Redclift, 1981; Dedousopoulos, 1985). Peasant families continue to produce both use-values for their own consumption (thereby securing part of their costs of reproduction) and exchange-values for the market. Since unpaid domestic labour is not imputed at its real market price or opportunity cost, the commodity price can fall below the level necessary to ensure that capitalist producers achieve the long-run average rate of profit. The result is a loss of surplus labour by peasant households and increased transfer of absolute surplus labour to other sectors and regions.

As these terms of production and exchange deteriorate and taking into account that prices of most agricultural products are centrally decided by the state, peasant producers attempt to restore their previous position by working longer hours and intensifying the use of unpaid family labour (Vergopoulos, 1975). The end result is an increase in the production and transfer of absolute surplus labour. Psichogios (1984) argues for Greek agriculture that from 1860 to 1976 Greek peasant producers had almost double their annual working days per holding (see Table 10.6).

Greek family producers, despite increasing mechanization which also increased their productivity, continue to work longer hours in 1977 than four generations ago. Taking also into account the decrease of family membership, we realize the pronounced intensification of work. Table 10.6 shows also that surplus labour in 1860 and 1939 was the same. Peasant families need to work a little more to ameliorate their living conditions. In 1976, however, necessary labour remained as in 1860, while surplus labour increased dramatically. Indeed, working longer hours by peasant families in rural regions was mainly for the benefit of other sectors and other regions. The household gained only its productivity increase. In effect, use-value production defines the margin by which peasant

Table 10.6: Working days and mechanization in Greek agriculture

	Annual working days spent by peasant families per holding				
	Total	Necessary Labour	Surplus Labour	Number of Tractors 1950 = 100	Fertilizers use (tons) 1938 = 100
1860	244	181 (1860)	63 (1860)	-	-
1939	253	190 (1939)	63 (1939)	-	100 (1938)
1953-60	286			100 (1950)	320 (1955)
1961-70	322			577 (1961)	1,000 (1966)
1971-77	340	180 (1976)	160 (1976)	3,550 (1976)	2,350 (1976)

Sources: Psichogios, 1984; National Statistical Service, 1951, 1961, 1980

families may be 'squeezed' through the deterioration in the prices of their marketed surplus.

Similar observations are made for Spain (Taragona area) by Garcia-Ramon (1982). She argues that new crops, such as irrigated hazelnuts, fruit trees, early potatoes and others, contrary to all expectations, require more hours of labour per hectare in 1974 than in 1955 despite a dramatic increase in mechanization (machinery, irrigation and fertilizers). She points out additionally that, if we add the increase in hours of the new crops to the decrease in permanent workers, we can see the considerable increase in the hours worked by the owner and his family, a pronounced intensification of work.

This also concerns the dependence of small peasant producers on commercial capital and state agencies for loans and their resultant increasing indebtedness. The loans and agriculture debt in reality is an additional transfer mechanism through which an intensification of work, and in some cases a productivity increase, can take place without necessarily corresponding to an income increase. This argument can be supported with data from Italy.

Table 10.7 shows that productivity and loans are both increasing faster than net real incomes.

Table 10.7: Italy: Index of labour productivity, loans and incomes in agriculture

Agricultural labour productivity	Loans to peasants	Agricultural net real incomes
1946 = 100	1946 = 100	1946 = 100
1962 = 272	1962 = 295	1962 = 167
1972 = 352	1972 = 340	1972 = 210
1975 = 381	1978 = 378	1975 = 230

Source: INEA, 1979

These problems are aggravated by the fact that many small peasant producers in southern Europe specialize in the so-called 'short-cycle, soft' agricultural products (certain fruits, vegetables, wine, olive oil and flowers) as opposed to 'hard' products (livestock, grassland, cereals, industrial crops, etc.) (see Maps 8.1, 8.2 and 8.3). Due to soft products' dependence on price fluctuations, sensitivity to spoilage and weather, small peasant producers are 'exploitable' by a large number of 'middlemen'. These middlemen, under state toleration, control harvest and distribution and often impose on peasants single-crop cultivation programmes. The result is frequently an increase in the prices to the consumer without corresponding benefits to the producer.

Another interesting relationship between agriculture and industrial capital which also supports the case of unequal trade and unequal rewards to labour is the one between small peasant producers (in traditional or dynamic farms) and independent farmworkers with agro-industries. Since the mid-1960s, a growing number of agro-industries established special contracts with small peasant producers in southern Europe and became the exclusive buyers of their products (Zalider, 1972). These agro-industries assign a specific cultivation programme to individual farmers, changing the traditional pattern of production for the local market. International firms, such as Nestlé in Spain (Galicia) and Unilever in Italy, and national firms, such as the Greek Tobacco Industries, the Greek State Sugar Company, the Catalan Vineries, the Andalusian Can Industries Company, the Southern Italian Tomato Company, and others, prefer to have this type of long-range contract with small peasant producers instead of owning and operating the land themselves.[6]

Taking the Greek sugar industries as an example, the whole system is developed in a pyramid-like hierarchy, at the top of which is the processing agro-industry and at the bottom the small peasant producer. The firm has a direct and indirect relationship with the peasant, either via subcontractors on a regional level who are responsible for the collection of products through local retail centres, or via direct contracts with individual farmers in cases of geographical proximity. The system is very efficient in allocating losses, production costs and reproduction costs among the small peasant producers; while at the same time protecting agro-monopolies from various problems such as high operating costs, weather conditions, labour costs, unionization and rural strikes.

Additionally, through this system, the big firm maximizes the extraction of absolute surplus labour from the bottom of the hierarchy without being directly dependent on its labour force: its 'workers' remain in name individual producers, while in reality they are reduced to a status of wage earners doing the equivalent of piece-work.

These observations on the three major channels of GTV can be further illustrated when we compare the GAI-AGVA index (see Chapter 9) for each region or province of the three countries with: (a) the geographical distribution of small peasant producers (percentage of cultivated area owned or operated by small peasant producers - those owning less than 5 hectares - in total cultivated area); (b) the geographical distribution of the organic composition of capital in agriculture (number of tractors per Km^2 of cultivated area, percentage of irrigated area to total cultivated area); and (c) the geographical distribution of agricultural wages compared with the national average (see Maps 10.1 to 10.15).[7]

Looking at Table 9.1 and at Maps 10.6 to 10.10 for Italy, we can see that the five top value-gaining regions (Emilia-Romagna, Piemonte, Lombardia in the north, Lazio in the centre and Campagnia in the south) present high figures in agricultural wages, tractors per cultivated Km^2, and extent of irrigated land and with the exception of Campagnia are specialized in 'hard' products. Emilia-Romagna and Piemonte have low percentages of small peasant producers, while the other three have high percentages. This is in sharp contrast with the top five value-losing regions (Marche and Toscana in the centre, Molize, Sardegna and Calabria in the south), which have high percentages of small peasant producers but at the same time low to very low figures for agricultural wages, tractors and irrigation and are specialized in 'soft' products. In addition, as Ceccareli et al. (1978) argue, transportation and communication infrastructure, housing and health facilities, banking and commercial services are far better organized in northern and central regions. Furthermore, the metropolitan area of Naples provides a large urban market for products and services inside Campagnia (Belli and Flora, 1978). Thus the pattern in Italy of value-losing regions in agriculture seems to follow the hypothesized combination of traditional capitalist and traditional small peasant production, with low organic composition of capital, lower wages and less developed intrastructure.

The pattern in Spain is somewhat different. The five top value-gaining provinces in agriculture (Leon, Lerida, Burgos, Logrono and Navarra, all in the north) have high percentages of small dynamic producers, with high figures in wages, tractors and irrigation and are mostly specialized in 'hard' products (see Table 9.2 and Maps 10.1 to 10.5). Among the top five value-losing provinces (Huelva, Cadiz and Sevilla in the south, and Oviedo and Lugo in the north) the first three have relatively low percentages of small farmers and the last two very high percentages. All five value-losing provinces, however, have low to very low figures in wages, tractors and irrigation. The difference here is clear between modernized and capital-intensive minifundia operated by dynamic farmers in certain northern provinces on the one hand, and backward latifundia in the south, and non capital-intensive minifundia in the north. A similar conclusion has been reached from a different perspective by other studies as well (see Malefakis, 1970; OECD, 1977). Additionally, the study of Camilleri et al. (1977) provides evidence about growing infrastructural differences between northern and southern Spain which influence the regional differentiation in organic composition of capital and wages.

In the case of Greece (see Table 9.3 and Maps

Map 10.1
SPAIN: Number of Farmers per Km² cultivated area (1975)

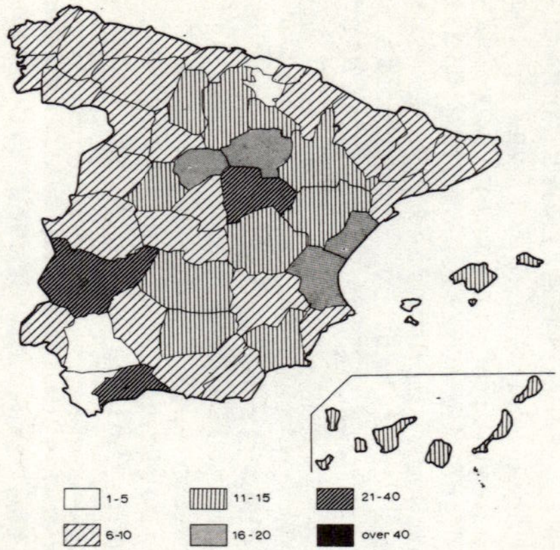

sources: Annuario de Estadistica Agraria, 1977; Banco de Bilbao, 1975

Map 10.2
SPAIN: Percentage of cultivated area owned or operated by small peasant producers (less than 10 hectares) in total cultivated area

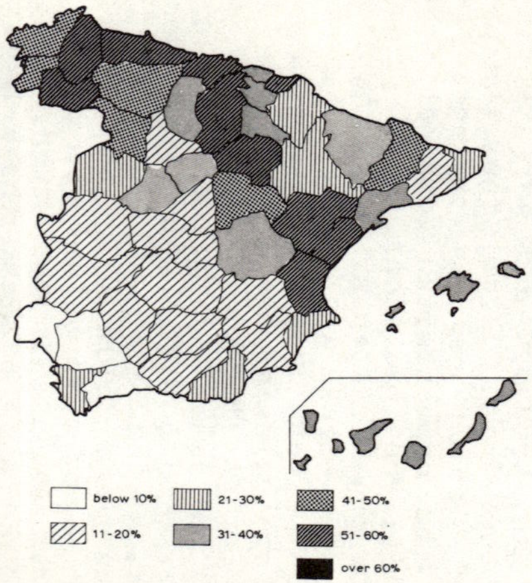

sources: Annuario de Estadistica Agraria, 1977, Banco de Bilbao, 1975

Map 10.3
SPAIN: Agricultural wages (1975, Spain = 100)

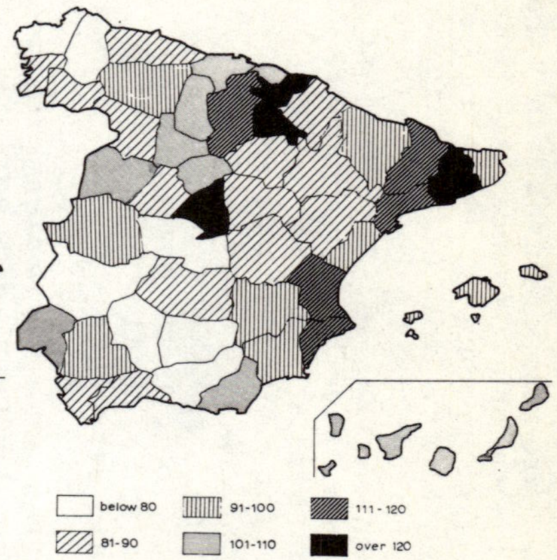

sources: Annuario de Estadistica Agraria, 1977, Banco de Bilbao, 1975

Map 10.4
SPAIN: Number of tractors per Km² cultivated area (1974)

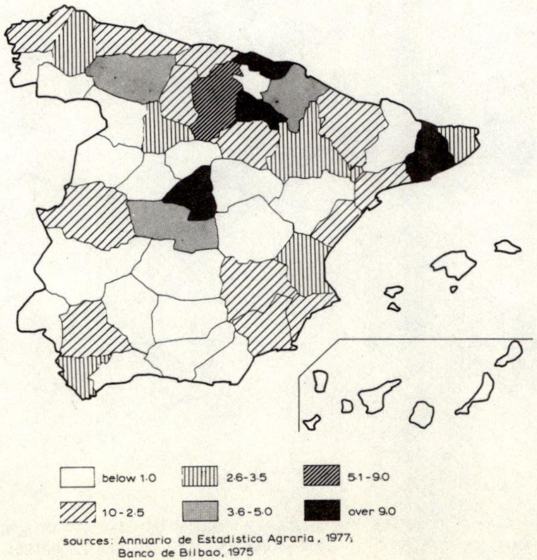

Map 10.5
SPAIN: Percentage of irrigated area in total cultivated area (1975)

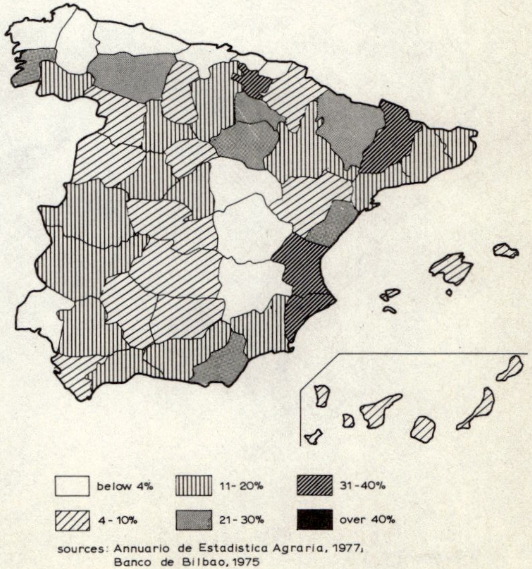

Map 10.6
ITALY: Number of farmers per Km² cultivated area (1975)

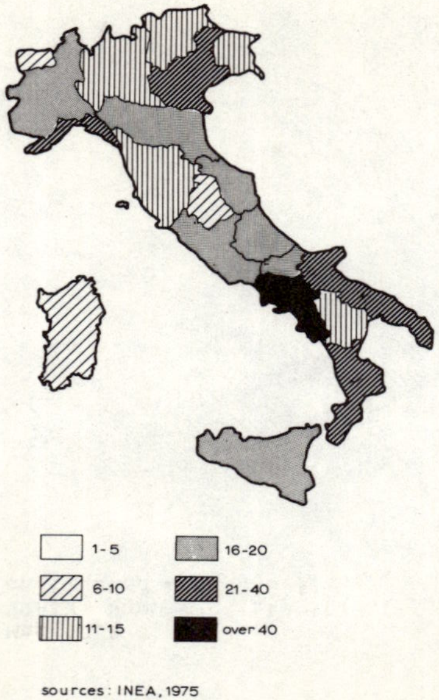

Map 10.7
ITALY: Percentage of cultivated area owned or operated by small peasant producers (less than 5 hectares) in total cultivated area (1973)

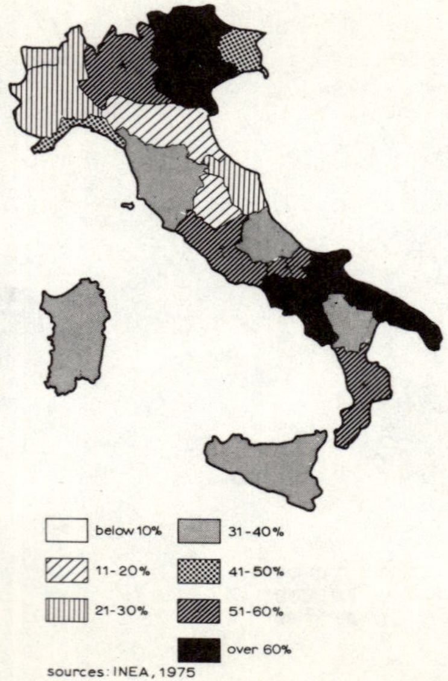

Map 10.8
ITALY: Agricultural wages (1975, Italy = 100)

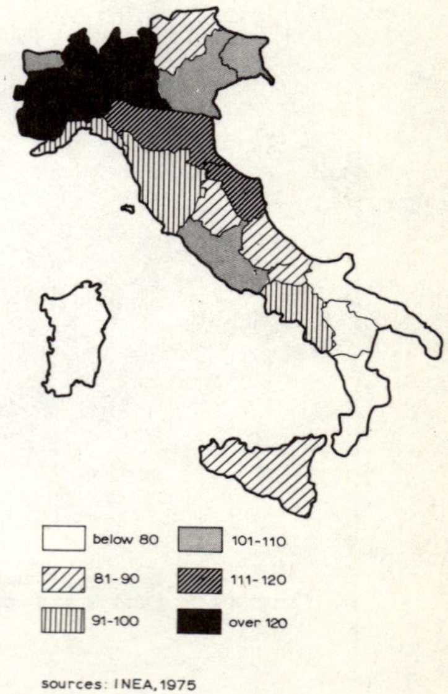

Map 10.9
ITALY: Number of tractors per Km²
cultivated area (1975)

Map 10.10
ITALY: Percentage of irrigated area in
total cultivated area (1974)

Map 10.11
GREECE: Number of farmers per Km² cultivated area (1974)

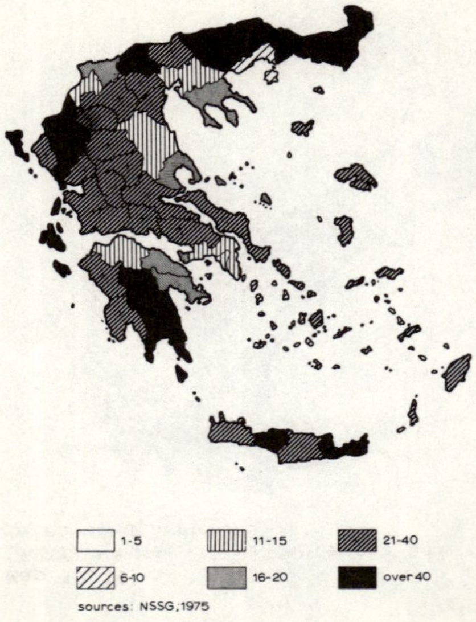

Map 10.12
GREECE: Percentage of cultivated area owned or operated by small peasant producers (less than 5 hectares) in total cultivated area (1974)

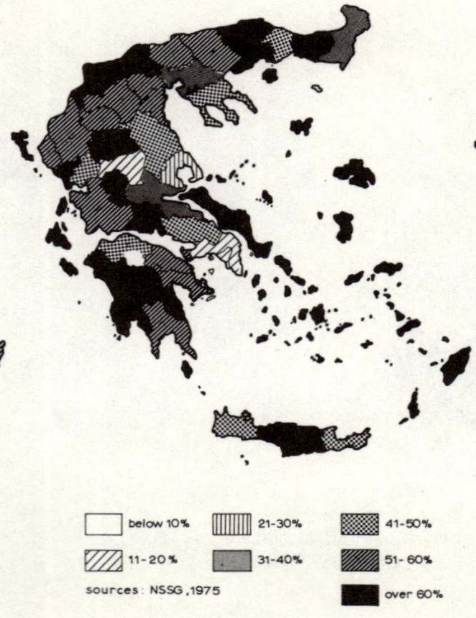

Map 10.13
GREECE: Agricultural wages (1975, Greece = 100)

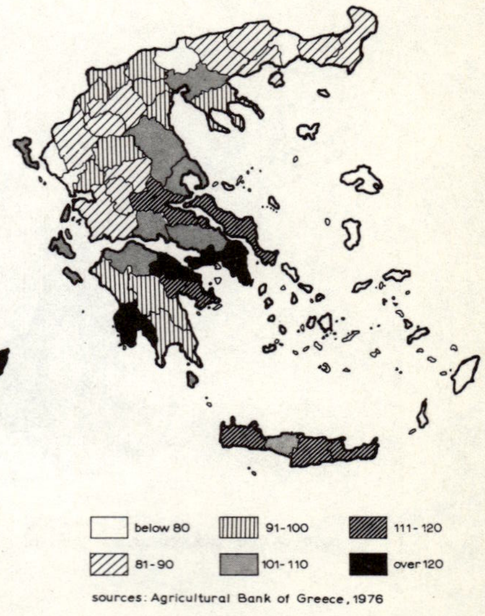

Map 10.14
GREECE: Number of tractors per Km^2 cultivated area (1974)

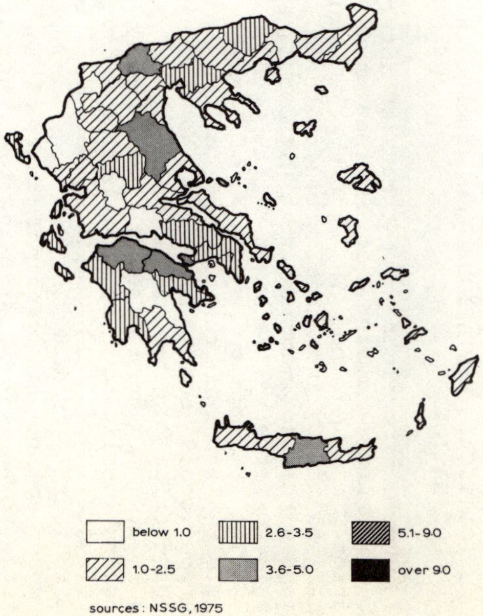

Map 10.15
GREECE: Percentage of irrigated area in total cultivated area (1974)

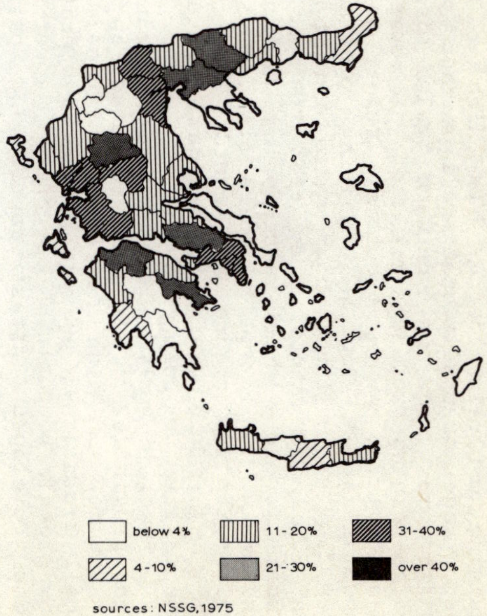

10.11 to 10.15) the high 'parcellization' of land does not leave room for latifundia-minifundia differences. With the exception of Thessaloniki and Athens, most of the top provinces, both value-gaining and value-losing, have high to very high percentages of traditional small peasant producers. Differences, however, between value-losing and value-gaining provinces are clear in terms of land-labour ratio, wages, tractors, irrigation and crop specialization. Top value-gaining provinces have low figures in land/labour ratio and high figures in other indicators; they tend to specialize in 'soft' products while value-losing provinces have the opposite. Exceptions here are Athens, with a relatively low figure for tractors per cultivated Km^2, and Iraklion, with low irrigation figures, both specializing in soft products.[8]

Thus, in Italy and Greece, value-losing regions and provinces in agriculture have an economic structure, with high land/labour ratios, based on small traditional peasant producers operating with low organic composition of capital, lower wages and less developed infrastructure. In Spain value-losing provinces also include traditional backward latifundia units. In terms of value-gaining provinces, in Spain and Greece their economic structure is based on modernized and capital-intensive small peasant producers, while in Italy and Spain there are also value-gaining regions with capital-intensive large-scale farming (traditional capitalist).

10.3 POLITICAL MEDIATION AND IDEOLOGICAL/CULTURAL DOMINATION

If anything is clear in the Marxist interpretation of agrarian evolution, it is that such evolution cannot be understood in strictly economic terms: the political and ideological/cultural dimension plays a role so important that without it one cannot begin to perceive the heart of the problem.[9]

The economic system of small peasant exploitation that contributed to GTV in southern Europe could not function alone to secure the reproduction of the social relations of production in the countryside. The necessity of institutionalizing and co-opting revolutionary attitudes and integrating various forms of agricultural production into the advanced market economy became a principal task of the bourgeois state, whose role in assuring the continuity of the system thus became fundamental. While the contradictions of the economy are such as to move the system towards equalization of the conditions of production and exchange, the reproduction of social relations in the countryside and the concrete conditions of regional underdevelopment generated tendencies towards differentiation. The whole system thus appears unstable, increasingly demanding a control applied through political mediation and ideological/cultural domination of the rural masses.

If economic domination leading to under-development (regional or national) is a process whereby the penetration of capitalism into peripheral regions mobilizes and then misallocates indigenous economic resources (i.e. allocates them in a way which benefits the core through direct and indirect GTV) then political and cultural domination can be seen as a process whereby the core's institutions mobilize local ideological resources (for political support, legitimation, control of representation) for the maintenance of the *status quo*, i.e. for the reproduction of the conditions of economic domination (Habermas, 1975). This, however, is not a pure 'external' intervention and can never be possible without an articulation and collaboration between core institutions and local social classes. Indeed, internal and external factors are dialectically interrelated, and attempts at establishing the superiority of one over the other in explaining political and cultural domination are superfluous.

A similar type of analysis permitted Gramsci (1975) to argue about Piedmontese hegemony over southern Italy, ensuring the political and cultural 'leadership' of north over south. This political and cultural subordination of the Mezzogiorno, according to Gramsci and later according to Mingione (1978) and Mottura (1980), was the *sine qua non* for the economic exploitation of southern peasants and consequently for differentially localized accumulation among the Italian regions. Similar observations are made for Spain by Bolloten (1968) and Malefakis (1970), and for Greece by Moskof (1974) and Tsoukalas (1975).

Looking at this process from the point of view of class analysis, one can argue that in a semi-peripheral political system (i.e. in the social formations of Spain, southern Italy and Greece) the bourgeoisie deal with the mobilization of the rural and urban working classes in a way which prevents their entrance into politics as *autonomous* political forces (Poulantzas, 1975; Mouzelis, 1978). The forces and the relations of production in developed capitalism provide a framework which makes possible (although, of course, not certain) the autonomous organization of the working class; whereas in semi-peripheral societies such as in southern Europe the setting does not leave much room for such an autonomous political solution.

The weaker southern European bourgeoisie could not apply social-democratic principles in politics as did its north-western European counterpart, because this threatened its own existence from a more radicalized and less willing to collaborate popular movement. Thus, as Poulantzas (1975) argues, the southern European bourgeoisie historically developed two broad political alternatives. The first was to abolish parliamentary institutions and through dictatorial means to depoliticize the rural and urban masses and exclude them from active politics. This has been applied on various occasions in southern Europe; from Italian and Spanish Fascism, to the 1967 Greek dictatorship. Under these circumstances the political mediation in the countryside was expressed through both a limited populism that partially satisfied the demands of small peasants, petty-traders and land speculators and relied on their support; and a defence of the interests of the big agrarian bourgeoisie and agro-industry through direct and brutal oppression of recalcitrant rural masses.

The second major political alternative was to use parliamentary and non-parliamentary institutions (e.g. the educational system, religion, police, the army) in such a way that the rural and urban masses were brought into the political arena in a dependent manner (Tsoukalas, 1975). This alternative produced a more sophisticated system for the reproduction of social relations in the countryside. It has to be seen, however, in a close relationship with the dictatorial alternative: the first is derived from the second and vice versa. In fact the development of capitalism in southern Europe occurred under authoritarian right-wing regimes which used the two alternatives according to the internal balance of forces.[10] Apart from the direct oppression during dictatorial regimes, the second alternative has historically taken various forms, among the most important of which are: political clientelism, exploitation of language and dialect differences, the institutionalization of share-cropping and the manipulation of agrarian reform measures.

1. Political clientelism has been a very effective system in southern Europe to control and integrate the peasantry into the central institutions of the state.[11] Clientelism must be seen as part of a process where the satisfaction of a common need will be viewed as a matter of personal privilege granted to the single client. In a situation of great scarcity, as was rural southern Europe until the mid 1960s, 'exclusive' access to the leaders was actively sought. Pressure groups and patronage agencies are strongly related to what Gramsci has called the 'rural intellectuals', whose role, according to Gramsci, was essential for the political and cultural domination of the peasantry. He wrote for Italy in 1929:

> (Intellectuals of the rural type) ... bring into contact the peasant masses with the local and state administration. Because of this activity they have an important politico-social function, since professional mediation is difficult to separate from political ... One can understand nothing of the collective life of the peasantry and of the germs and ferments of development which exist within it, if one does not take into consideration and examine concretely and in depth this effective subordination to the intellectuals. (Gramsci, 1971, Prison Notebooks, p.14)

A good example of Gramsci's observations is the southern Italian 'pettifogging lawyer' (*paglietta*), who during the 1920s ensured contact between the rural proletariat, small peasant producers, big landowners and the state apparatus. Northern hegemony over the Mezzogiorno was maintained through favours to this intellectual stratum in the form of jobs in the public administration or licence to pillage the local administration with impunity. Thus, the social stratum which could have organized the endemic southern discontent instead became an instrument of northern policy. This arrangement became particularly important after World War II, when the south ceased to be simply a region from which resouces were drained and has become in recent years a region that absorbs an influx of resources, both in the form of investments and in the form of a flow of public spending for salaries, pensions and so on (Pinnaro, Pugliese, 1975). This profound alteration in the region's economic role corresponds to an equally important change in the nature of clientist relations. A new 'mediator' class appears on stage: the state petty bourgeoisie. Its new role is to keep social tensions in rural areas at least partly under control and to replace direct class conflicts by other, more mediated, ones. It is not 'bosses and workers' any more, but 'local government authorities and the marginal population' (Ginatempo, 1985). The farm-worker's position thus develops into what Pugliese (1983) calls a 'worker on state benefits, a precarious client of the welfare state'. He points out that in southern regions a large number of the rural population survives by making a living through employment in public works, mainly in reforestation. This type of employment and its allocation process (through personal favours and exclusions) lead to new forms of social control and reproduce ties of clientelism to the central state and local authorities.

In Greece, each family in a village attempts to establish an exclusive and particular relationship with persons in power. Decision-making or initiative is seldom possible at the lower levels of the state bureaucracy because of the reluctance of central officials to delegate power. To reach that level at which their affairs can be effectively dealt with, small peasant producers are dependent on a system of patronage (Campbell, 1974; Mouzelis, 1978). Patronage is the means through which the local community is linked to the wider social structure. And it is significant that the group which seeks this connection is usually the individual family of small peasants and not the individual or the community.

Local patrons - mainly lawyers, merchants or doctors - have direct connections with persons of influence in big cities and through members of parliament are linked to the legislative assembly. Local patrons act as political representatives (Greek: komatarchis) responsible for the collection of votes through personal favours - and, on occasion, blackmail. In this respect, the organization of government and the structure of patronage in the countryside are parallel hierarchies. At any particular level the two hierarchies are related through ties of domination and dependence (Tsoukalas, 1975). Since World War II and the Civil War, right wing governments effectively managed the patronage system for their electorcal victories. Until 1975 (the year of the first post-Junta elections) an atmosphere of mistrust, fear and oppression was predominant in rural areas. After 1981, when PASOK came to power, only minor changes took place. The rightist straightforward patronage system changed to a more sophisticated one. The rural masses, now typically 'participate' in local administrative bodies, but decisions have already been made elsewhere. For example, district councils - an important new institution - instead of being elected, are appointed by PASOK and chosen on the basis of party membership. Thus, rural population is still largely excluded from direct participation in local decision-making and government hierarchy, but through the new patronage system can enter policies in a dependent manner.

Similar observations can be made for Spain, where clientelism and patronage in the countryside before and during Francoism was known as <u>çaciquismo</u> (Lorenzo, 1969).[12] As several studies have noted, big bosses in rural areas controlled votes on a family basis, connecting agricultural economic transactions (e.g. loans, purchase of products) with voting for specific candidates. During the Opus Dei period, the pursuit of prosperity and modernization - '<u>desarrollismo</u>' - modernized this patronage system through the channels of local and sectoral development agencies. The latter continued after Franco's death. Apathy and conservatism among the rural population in post-Franco Spain may thus be related to a continuing deep distrust of politicians on all levels of government. Small farmers have low expectations for any real improvement in local conditions regardless of political promises and therefore prefer the candidates they at least know.[13] Under such conditions it is easy for traditional bosses to manipulate the vote by circulating rumours and employing direct and indirect patronage links.

Another type of rural intellectual in Spain, now beginning to disappear, was a section of the clergy tied to big rural bosses (usually big landowners) referred to as <u>beneficiado</u>. The beneficiado was a priest who was supported by a wealthy person to do works for the faith (Hansen, 1977). Among others, a duty of the beneficiado was to visit the small peasants of the area and to report back to the powerful local patron about the peasant's political adherence, his opinion of the patron and his family's moral qualities. The spy system led to the exclusion of many share-croppers from their land. It also turned many of the peasantry into ferocious anti-clericalists during the Civil War.

2. Another form of cultural domination over the peasantry is the exploitation by the state and the bourgeoisie of language and educational differences (De Vos, 1975); Kiray, 1973; Tsoukalas, 1975). Peasants in the three countries suffered during the period under study from an ideologically manipulated language dichotomy: in Italy and Spain a split between the official national language and regional languages and dialects (Basque country, Catalonia, Andalusia, Sicily, Napoli, Veneto and others) and in Greece a split between Katharevoussa (the pure archaic language) and Demotiki (the popular living language). This cultural domination established itself historically, with the development of the central state. The exploitation of the language dichotomy by an educated élite tied with the state apparatus made the patronage system the only path through which the illiterate small peasant could approach the state bureaucracy. Thus the reproduction of dependency of small farmers to their patrons has been secured through the inadequate education of the rural masses and the establishment of an 'official' language almost foreign to them.

The unification of Italy and Spain and the corresponding process of state development in Greece coincided with a concentration of political and cultural power in certain urbanized regions and the reduction of rural areas to a status of peripheral regions of low cultural prestige (Gramsci, 1975; Treves, 1965). Peasants found themselves under ideological pressure to change their local language and local mode of expression through the official educational system imposed by the state and the mass media. Thus, the antithesis between city and countryside developed a strong political and cultural/ideological component with obvious territorial consequences: a system of core regions, where a dominant culture group extends its influence downwards to culturally subordinated groups; and a system of peripheral regions, where groups of low cultural prestige contribute to maintain political stability and legitimation of their subordination, so long as they consume the cultural stereotypes 'exported' by the core.[14]

In Italy the new Italian language of the national state, which developed after the Risorgimento, has been resisted by various 'mother tongues' in many Italian regions despite the development of the mass media and state education since World War II. But this has had little effect. If political incorporation resulted in peripheral economic dependence, it simultaneously gave 'northern' cultural forms superordinate status within southern societies (Mingione, 1976). Indigenous élites in these regions which sought the increased opportunities afforded by the Risorgimento and later by the process of industrialization began to assimilate by learning other languages. The process has been called by some Italian Marxists 'Piemontetion' of the south, referring to increasing acculturation of southern groups and the adoption of northern values, language, norms and life-styles (Libertini, 1973; Donolo, 1972).

In Spain, the Castillian bourgeoisie also attempted to promote its values, religion and language over the rest of Spain, but it succeeded only partially in the south (Vicens, 1969). The so-called 'Castillization' process, designed to preclude the possibility of regional threats to Castillian hegemony, has been defeated in the north, where the Galicians, the Basques and the Catalans successfully defended their languages (Hansen, 1977). It has succeeded, however, in dividing Spain into various regional cultures opposing Castillian domination, which now after the death of Franco are the basis for strong movements for regional autonomy.

In Greece, the split between Katharevoussa and Demotiki has been more important than regional languages, such as Turkish in eastern and central Macedonia and a mixture of Slavic dialects near the border with Albania, Yugoslavia and Bulgaria (Eliou, 1978). Since 1922 and especially after the Civil War, the Greek state, following chauvinistic plans, destroyed - often brutally - the many local languages, considering all dialects as 'against Hellenism'.[15] The Greek educational system, as Tsoukalas (1975) argues, contributed to the reproduction of Katharevoussa and through this to the cultural domination of the rural and urban masses. In fact, in the Greek case, the identification of cultural domination with class domination developed under the guidance of the state is clearer than in Italy and Spain (Poulantzas, 1975).

3. A third example of political and ideological mediation was the institutionalization of long-term share-cropping especially in regions such as southern and northern Spain, the Mezzogiorno, Sicily

and Sardegna, the Peloponnese, Crete and the Aegean Islands. In these regions, share-cropping made possible the reproduction of the small peasant economy, even before agrarian reform measures.

In northern Spain and in regions such as Galicia, share-cropping contracts were called *caseiro*, while in Catalonia, where intensive viniculture is widespread, share-cropping contracts took the form of the *rabassa morta* which was designed to settle cultivators on the owner's land for a long period (Hansen, 1977).[16] The *rabassaire* system, institutionalized since 1880, had an important political role in Spain: in contrast to the miserable wages of the unskilled rural and urban proletariat, a *rabassaire* had a great deal of influence over his own earnings and consequently more latitude in life horizons than did the man who settled in the proletarian ghettos of big cities. The chief virtue of the institutionalization of share-cropping, as Mintz (1970) has pointed out, was '... to avoid the proletarianization of the countryside'. This was extremely important until the period of the Civil War for the Catalan and Basque bourgeoisie, which were involved in two kinds of conflicts: class conflict with the proletariat within the region, and political conflict with the central state for autonomy. The avoidance of proletarianization of the countryside was also helpful during the Franco dictatorship in that it helped develop an atmosphere of relatively stable class relations until the 1970s. In 1975, 65 per cent of the agricultural land in Catalonia, 58 per cent in Galicia and 62 per cent in Andalusia was still worked by share-croppers (Moreno and Rodriguez, 1979).

In the Mezzogiorno, share-cropping is called *mezzandria* and is closely linked with absentee landlords. This means that there exists, in proportion to gross landed income, a large group of petty and middle bourgeoisie living on 'pensions' and 'rents', and typically producing the main opposition to agrarian reform in the south (Saraceno, 1959). Share-cropping in the south was institutionalized during Fascism, forcing the rural population to stay on the land. Since World War II, the Italian government has tried to change these rules and in 1964 declared a special law to do so but with only limited subsequent success (Ministerio di Agricultura, 1974, p.86).

A similar type of share-cropping has been developed in the Peloponnese, called *sebria*; and in the Aegean Islands, called misiaka, introduced from Italy (Moskof, 1974). Both systems were institutionalized at the turn of the century and are still in operation. In the case of the Peloponnese and Crete, Tsoukalas (1975) argues that the system succeeded in the elimination of anarcho-syndicalist movements. At the same time it contributed to out-migration because the contract provided the owner with the power to exclude the share-cropper in cases of bankruptcy.

4. Finally, a major example of state involvement for the reproduction of small peasant producers has been agrarian reform. While agrarian reform is basically an economic intervention by the state, its political and ideological implications are of major importance. It has been argued that agrarian reform typically contributes to capitalist development by putting an end to the rents that are paid to 'parasitic' large landowners, who by their mere possession of large amounts of land are able to obtain super profits (from capitalist tenants, peasant share-croppers, price control, etc.; see Lenin, 1956; Chayanov, 1966; Goodman and Redclift, 1981). Through the elimination of latifundia via agrarian reform measures, surplus value and surplus labour that previously were directed to ground rent can be transferred to other sectors and regions of the economy and utilized especially for the expansion of urban-based industrial capitalism. This has been the focal point of the class conflict between the landed aristocracy and the industrial bourgeoisie in Europe since the 18th century and was an important surce for the differences between north-western and southern European agrarian structures, as well as for the differences among southern European regions.

In north-western Europe, the industrial part of the bourgeoisie was powerful enough to eliminate effectively large land ownership. Industrialization meant either the destruction of simple commodity production in the countryside or the articulation into the CMP of small peasant producers (resulting from agrarian reforms) in such a way that they managed through modernization and specialization to increase their productivity and to establish a 'positive complementarity' with the rest of the economy (Rey, 1973). This was made possible, as mentioned earlier, through political institutions based on social-democratic principles and the willingness of working class organizations to collaborate with the state.

In southern Europe, the weaker industrial

bourgeoisie, in order to defend its class interests against a more antagonistically politicized working class, found it necessary to make an alliance with the landed aristocracy. The working class and the peasantry have never really had a stable share in political power, since this threatened capital (as shown by civil wars and the repeated short-lived experiments of popular front movements). In addition institutions in southern Europe functioned mainly to institutionalize popular and peasant participation in state institutions via patronage. The popular masses lost their power, and such 'participation' as there was degenerated into a bureaucratic system which in some measure obtained the support of poor masses and manipulated the situation to save the interests of the classes in power (Vergopoulos, 1975).

Agrarian reform has taken place in southern Europe only when the bourgeoisie felt sufficiently strong to risk sacrificing its landowner fraction: in Greece from 1897 to 1922, in Spain a slow process from 1932 and in Italy since 1954. Land redistribution has come about not as a result of political and social strength of the rural masses, but as a result of the strength of the capitalist class (Rey, 1973). Petty commodity production remained relatively stagnant and technologically backward compared to north-western Europe.

Thus, agrarian reform in southern Europe contributed to the development of capitalism mainly in political and ideological terms. It made it possible for the state to co-opt the political support of the peasantry and control their reaction to the expansion of capitalist relations of production in the countryside (Mouzelis, 1978). It played a double role in destroying traditional peasant communities and avoiding the militant proletarianization of the peasantry, by creating the basis for the development of a new class of small peasant producers tied to the capitalist market. In this respect the creation and subordination of small peasant producers had important political and ideological effects. Remaining in name only the owners of their means of production, small peasants supported bourgeois policies and they came to believe that their interests significantly diverged from those of the (urban) working class. This was the main barrier which, until recently, hindered their participation in common struggles.[17]

In summary, I have attempted here to formulate an interpretation of the GTV taking place in the agricultural sector. The interpretation was based on the analysis of the articulation of various forms of capitalist and peasant production and the capitalist mode of production. As noted in Chapter 7, this is important since we are dealing with social formations which include sectors and regions operating under less capitalized forms of production. A primary conclusion is that the articulation between these forms and the CMP is concretized simultaneously at economic, political and cultural/ideological levels. The analysis of 'concrete capitalism' through the example of agriculture in the three countries has shown that DLA and GTV do take place at the economic level, but the spatial division of labour is concretized only after the interrelation and mediation of political and ideological/cultural instances. Thus, any analysis of uneven regional development in southern Europe and consequently any action dealing with it must take into account the multidimensional character of the phenomenon which points directly to the role of social agents including the modern capitalist state.

NOTES

[1] See, Agricultural Bank of Greece (1975) and B. Kayser et al., (1971).

[2] A number of local reactions against those investments in the Peloponnese took place in Ilia region during spring 1978, see newspaper AYGI, 12 April 1978.

[3] The problem of whether an agricultural labour market exists is addressed by Mottura (1980) and Paci (1979). Both agree that regional characteristics determine the degree to which a labour market exists.

[4] Kavouriaris (1974) estimated that Greece 'exported' in 1970 through its labour force to West Germany a value equivalent to US$896,478,533 (in 1970 prices). During the same year remittances from West Germany were only US$343,000,000.

[5] Terms of trade against agriculture as a transfer mechanism is pointed out also in several studies outside southern Europe. T.H. Lee's (1971) study of Taiwan, while using domestic base period prices instead of international prices as the point of comparison, indicates that the deterioration of agriculture's terms of trade in the 1950s and 1960s as compared with the 1930s involved an implicit (or, as he says, 'invisible') transfer from agriculture

to other sectors equivalent to one-half to two-thirds of the real capital outflow from agriculture. Studies in Pakistan indicated that in the 1950s perhaps as much as 10 per cent to 15 per cent of agricultural income was being transferred out due to adverse terms of trade relative to world prices (Lewis, 1973). Little, Scitovsky and Scott, 1970, by recalculating the GDP of 10 developing countries at international, instead of domestic relative, prices, found that agriculture was subsidizing industry by 10 to 20 per cent or more of agricultural value-added as a result of trade policies. Finally, De Janvry (1981) found for Chile that agriculture as a whole gained as much as 15 per cent of its value added from changes in relative prices from 1959/1961 to 1962/1964, which suggests a similar loss of income in the earlier period relative to the later one.

[6] For a similar type of relationship in Latin America, Harris pointed out that:

> Another important characteristic of contemporary capitalist development in rural Latin America is the existence of a special relationship between capitalist agricultural enterprises and smallholding producers. Luisa Pare has analysed one example of this phenomenon among the *caneros* in Mexico. These smallholding peasants produce sugar cane on their own land, but they are financed and organized by private or state capitalist sugar mills. They sell the sugar cane they produce to the capitalists who finance them and who give them only enough remuneration to reproduce their labour power. (Harris, 1978, pp.9-10)

[7] All data to construct these maps were drawn from official statistics of the three countries. For Italy and Greece there were also original maps available, except for wages. For Spain smallholdings are considered below 10 hectares. I have not found comparable data for the construction of an infrastructural index on a regional/provincial basis. For infrastructure, I refer to secondary sources and case studies for each country.

[8] See, Agricultural Bank of Greece (1978), pp.86-9, for a discussion of the infrastructural and other contrasts between urban and rural provinces and the inadequacies of the transport and communications networks.

[9] Such an approach goes back to the complex relationship between urban and rural social classes

and the fundamental antithesis between city and countryside. Its systematic explanation is an extremely complex affair requiring analysis of the genesis and development of southern European institutions in the context of the class struggles which can ultimately account for them. Such an enterprise, however, is beyond the scope of this work. All that can be presented here are some suggestions and a discussion of a few examples indicating the areas where one might look for answers.

[10] However, one must be careful not to go to an extreme and identify the 1967 Greek dictatorship with right wing totalitarian regimes. It is important to see the differences between the Greek dictatorship and Fascist or even quasi-Fascist regimes, as was the case in Spain and Portugal. Neither Papadopoulos nor his short-lived successors managed in any serious way to build up totalitarian organizations for mass mobilization and support on the pattern of Nazi Germany or Fascist Italy and Spain. See, Poulantzas, 1975; and Mouzelis, 1978.

[11] Clientelism has been studied mainly from a behavioural perspective and far less in its relation to the political system. For a good critical overview on clientelistic theories and their application to Greece, see Mouzelis (1978). Of all classes, the middle classes and small peasant producers seem to be more responsive to individualistic incentives and clientelistic practices. Marx himself, in the 18th Brumaire, notices that:

> The small peasants are consequently incapable of enforcing their class interests in their own name, whether through a parliament or through a convention. They cannot represent themselves, they must be represented. Their representative must at the same time appear as their master, as an authority over them, as an unlimited government power that protects them against the other classes. (Marx, 1963, p.124)

[12] A similar structure of political mediation has been described by Bartra (1975) for Mexico. As Bartra argues, the 'caciques' originated with la reforma agraria and obtain their position by means of kinship ties, friendship, debts, favours, threats, etc. In time, this system breaks down, but the cacique has transformed his power into wealth and exerts his control in a despotic and arbitrary manner. See, Bartra, 1975, pp.143-4.

[13] Votes for the centre-right Spanish government in 1978 came mainly from small peasant producers in central and southern Spain. See El Pais, 10 April 1979.

[14] This type of unequal relationship between core and periphery has been often called 'internal colonialism', better introduced in the regional context by M. Hechter's work (1975) focusing on Wales. While Hechter's book has some valuable contributions, the internal colony model is not an adequate basis upon which to build an understanding of the political economy of underdeveloped regions. For an extended review and critique of the theory, see, Cooke, 1983.

[15] In Greece there are no current official figures for minority populations. With all the reservations which the nature of the question naturally imposes, the following table is based on the 1951 census:

Greek citizens whose language is not Greek

	Number	Per 1,000 in total Greek population
Turkish	179,895	23.57
Slav	41,017	5.37
Koutsovlach	39,855	5.22
Albanian	22,736	2.98
Pomak	18,671	2.45
Armenian	8,990	1.18
Gypsy	7,429	0.97
Others	16,330	2.14

Sources: NSS, 1951; Eliou, 1978

Turks, Slavs and Koutsovlachs are mainly concentrated in the northern provinces, while other minorities were dispersed throughout the country.

[16] The rabassa morta system was a long-term share-cropping contract based on the life of a vineyard, terminating after two-thirds of the original vines had died. Shares of the harvest were one-third for the proprietor and two-thirds for the share-cropper (rabassaire). No money changed hands, and it was incumbent upon both parties to produce wine independently of each other. Finally, costs and

methods of cultivation were in the hands of the share-cropper. See, Vilar, 1962; and Hansen, 1977.

[17] Some common struggles did occur during the Spanish and Greek civil wars and the Resistance of Italy. Since the 1970s economic problems and the application of the EEC's agricultural policies made the southern European peasantry extremely sensitive in politics. The first Pan-European rural strike organized in June 1977 enjoyed the support of all major European trade unions. See, L'Unità, 6 December 1977.

Chapter 11
REGIONALIZATION AND REGIONALISM

Throughout this work special attention has been devoted to the political and ideological importance of spatial organization in the reproduction of capitalist social relations. This has been analysed in detail historically, by taking agriculture as an example, and with a direct discussion of the central and local state. The reproduction process of capitalist relations (including the reproduction of core-periphery regional relations) necessitates, at every stage of development of a social formation, the intervention of the state and local authorities, as putatively 'separate social agents', apart from individual capitalist interests. In what follows I will attempt first to discuss briefly contemporary political developments in southern Europe, focusing on three interrelated issues: clientelism, corporatism and authoritarianism; and secondly to discuss emerging regional mobilizations, questioning both the practices of the state and regional backwardness. These mobilizations, however ambiguous they may often be, have a directly political character and under a certain conjuncture may become major agents for social and spatial change.

11.1 THE POLITICAL BACKGROUND: CLIENTELISM AND AUTHORITARIANISM

Uneven regional development and the core-periphery structure in southern Europe were a concrete fact for a long time before the people in peripheral regions became conscious of it. On the one hand the masses in peripheral regions were too little politicized seriously to threaten the stability of the central state; while on the other, regional uneven development appeared 'legitimate', as a necessary and presumably temporary misfortune of economic progress. So long as the whole process was not associated with the role of the state and local authorities, there was no 'regional crisis' as such. The accumulation and distribution of surplus value among social classes and regions were necessary starting points, but they do not in themselves explain how a core-periphery relationship may generate a regional crisis. What we are talking about is in fact a multidimensional process. The degree to which a core-periphery relationship is developed towards a regional crisis within the contemporary capitalist state is a function of the degree to which the uneven distribution of the surplus product between two different regional bourgeoisies acquires increasingly political and ideological/cultural dimensions.[1] Thus, while uneven regional development is the outcome of an historical process of peripheralization and core creation, the regional crisis appears only under a certain conjuncture, when it becomes associated with political and ideological instances and more specifically with the modern capitalist state (Dulong, 1978; Hadjimichalis, 1983).

Since the end of the 1960s, in most of the countries of western Europe, intensified state control over every sphere of socio-economic life, combined with extensive spatial policies, gave birth to a new form of state domination, what Poulantzas (1978) called 'authoritarian statism'. According to Poulantzas, authoritarian statism corresponds to the current phase of advanced monopoly capitalism in the way that the liberal state was associated with the competitive state of capitalism (Poulantzas, 1978, p.204). Authoritarian statism involves neither a Fascist regime nor a totalitarian dictatorship. While under a certain conjuncture it could develop towards these forms, as it appears today authoritarian statism is both the outcome of increasing economic intervention and a reflection of the growing radicalization of wider segments of popular masses.

In southern Europe, however, authoritarian statism cannot be analysed along the same lines. It has acquired a number of distinct characteristics related to a weaker civil society (compared with northern Europe) and the much more widely applied practices of clientelism and patronage. With important regional exceptions and differences, already relatively weak civil societies were further subordinated by centralized and inefficient central state and other institutions such as the church and the army.

From the mid 1950s to the beginning of the 1970s clientelism and patronage were the major mechanisms for exclusion/inclusion of large parts of the population in the process of rapid economic growth. This is not to say that classes did not appear on stage as such, but it seems that there was not a one-to-one correspondence between class 'places' and political practices. In Italy, under DC hegemony, the practice of clientelism resulted in an institutionalized tripartite relationship between government agencies, private interests (e.g. Confidustria, Catholic Action) and the party machine. In addition, the large number of state agencies and parastate organizations (railroads, telecommunications, electricity, social security and so on) permitted the building of a vast patronage network which has been particularly important in consolidating DC rule in the south.

Similar developments have taken place in Greece. Clientelism acquired momentum during the pre-dictatorial period and continues today (with important interventions from mass mobilizations) especially through employment 'favours' in parastate organizations, loans provision for small-scale investments and through the allocation of public resources and public works. Under the rule of ND, clientelism 'privatized' politics, when various groups and interests (such as shipowner capital, commercial capital, hotel owners, developers, contractors and so on) had direct unmediated access to political authority, which they treated as a tool for their private aims.[2]

The political situation in Spain was somehow different. Francoism produced an entirely new economic situation and an expanded urban culture that could no longer be contained by the old authoritarian patronage form. As the Spanish bourgeoisie was looking for new space to manoeuvre, important Francoist groups (the Falange, landholders, the church) have lost popular support (Giner and Sevilla, 1984). These groups have been replaced gradually in their mediating roles by right wing technocrats of the Opus Dei, pursuing economic progress, and by local administrators.

In these countries, as we saw in previous chapters, contemporary state forms (since 1974) seem to be more 'open' and 'democratic' compared with the earlier experience of Fascism and military dictatorship. At the same time, however, the deep historical roots of clientelism and authoritarian rule in southern Europe permitted the development of a distinct form of authoritarian statism parallel with the development of formal but limited democratic liberties. The outcome has been a direct political crisis and a crisis of the state itself. Poulantzas, linking the crisis of the state with authoritarian statism in southern Europe argues that:

> ... in certain European countries we are witnessing a real political crisis which finds expression, moreover, in a crisis of the State. Hence, in *these countries* authoritarian statism is marked by a State crisis: to a varying degree, this is the case in Spain, Portugal and Greece, as well as in Italy and France. (Poulantzas, 1978, p.206, emphasis in the original)

Authoritarianism and clientelism through associated changes in the social and spatial division of labour appear to be leading towards a considerable shift in class relations in southern Europe. The period of rapid 'growth' during the 1950s and 1960s and changes in the labour process itself have deepened the social and spatial division of labour in the three countries (see, Secchi, 1977). Intensified exploitation has rested on more complex and refined forms of labour control, such as work speed-ups, higher labour productivity and degradation of living conditions through urban and regional segregation, inadequate means of collective consumption, etc. The world economic crisis expanding since the mid 1970s with inflation and selective unemployment, however, introduced a much slower process of economic growth, which had an immediate economic and social impact on the legitimacy of the state. Grand promises were never fulfilled, while constant shuffling of cabinet posts and leadership roles did little to change patronage networks. As an outcome, the specific form of southern European authoritarian statism was unable to act as a viable rationalizing force that could implement effective social reforms and overcome its crisis of legitimacy. As Ingrao (1976) points out in *Critica Marxista*, clientelism and authoritarianism can no longer act as a stabilizing force in southern Europe; on the contrary, it seems that it is itself an important factor of destabilization. The paradox lies in the fact that these practices are not simply the means with which the state equips itself to tackle the wider political and economic crisis, but the product of a deep structural crisis which the state itself helps to produce and reproduce (Poulantzas, 1978).

Authoritarianism as ideology and mode of administration is spread to local levels as well. Local authorities more and more command local resources and services as 'patrons', reducing citizens to passive recipients of services. Thus, the standard mode of service delivery usually involves a distinctive combination of low efficiency/low quality service provision with increasing paternalism and authoritarianism. This combination generates a series of deficiencies and contradictions that often makes the popular masses feel oppressed from institutions and policies which many times are the outcome of their own struggles (e.g. unemployment benefits, municipal hospitals). Thus, authoritarian statism (central and local) does not correspond to a univocal strengthening of the state and local authorities: it rather involves a contradiction of <u>strengthening and weakening</u>, given the various elements of crisis including the regional crisis.

All these developments have stimulated since the mid 1960s a rise and politicization of struggle expressed in the demand and the forms of mobilization found within southern European popular movements. One of the most important effects was the rise of a new popular awareness concerning questions that can no longer be considered secondary, such as struggles over cultural values, feminism, the student movement, the urban environment, collective consumption, and uneven regional development. At the same time, authoritarian statism also became the basis of a neo-conservative attack on welfare institutions and planning policies, aimed at reducing the importance of the state and to disencumbering the free market. While neo-conservatism appears at present to lack a strong social basis in southern Europe, the performance and policies of contemporary socialist governments will determine if such a basis will develop in the future.[3]

11.2 REGIONAL MOBILIZATIONS AND REGIONAL SOCIAL MOVEMENTS

Among the new forms of struggle which combine class interests with regional, cultural and community identity, regional mobilizations seem to play an increasingly important role. Regional mobilizations arise from both objective conditions of uneven regional development and subjective conditions of rising regional consciousness. There are, however, many forms of regional mobilization in southern Europe as well as conflicting explanations of their origins and importance.[4] One, for example, cannot put under the same umbrella the Basque ETA, a military movement based on nationalistic anti-capitalist principles; the Andalusian mobilization, a petty-bourgeois/small peasant coalition against the central state; the Calabrian movement, a leftist self-management/cooperative movement; and the Cretan movement, a petty-bourgeois integrationist regional mobilization. Their common 'attachment to a place', in other words their territorial base of existence, cannot alone define them as regional social movements.

By regional social movement I mean a certain type of organization of social practices at a regional level, the logic of whose development: (a) has a territorial and multiclass base; (b) contradicts the institutionally dominant social logic, i.e. it is against authoritarian statism and unregulated capital; and (c) challenges conditions of uneven regional development, i.e. it is against the distinction between aggregate economic growth and territorially adapted development. In this respect regional social movements constitute a critique of both the conventional model of capitalist development and the capitalist state. They are specific reflections of the growing antithesis between the deterritorialized state and the territorial community (Ledrut, 1975).

Regional mobilizations are outcomes of certain conjuncture when regional consciousness is built along the two interrelated dimensions of: (see also, Melucci, 1981; Castells, 1983; Rokkan and Urwin, 1983)

a. '<u>membership space</u>', that is membership in a group that has, and becomes <u>aware</u> of having, some common socio-cultural, economic or political problems and aspirations.
b. '<u>geographical space</u>', that is identification with and occupation or use of a specific territory.

These two interrelated dimensions of regional mobilizations gave to 'neutral' concepts such as territory, culture or economic inferiority a politically significant dimension because of the interpretation and value placed upon them by people. Under these circumstances, regional mobilizations could struggle against an adversary for the seizure and control of what the mobilizations identifies as valuable resources (e.g. local employment, preservation of local language, administrative autonomy, environmental protection, and so on).

In southern Europe these issues acquired special dimensions. First, recent experience of Fascism and military dictatorships left centralized administrative structures untouched. The latter made the demand of decentralization and self-government at local and regional levels a first priority towards the democratization of public life. Secondly, the deep structures of authoritarianism and clientelism combined with highly unequal patterns of regional development made backwardness a valuable conjuncture for regional consciousness building. Thirdly, cultural distinctiveness in certain regions implied some degree of consciousness of autonomy (or even separateness), where the impact of history upon identity was that of ensuring the retention of 'collective memories'. And fourthly, collective experience of civil wars, Fascism and military rule provided a double legacy: on the one hand a strong left (in size and as an ideological catalyst within each social formation) and on the other a popular, weak but politically influential extreme right. These factors have deeply influenced regional mobilizations in southern Europe and differentiated them from movements in other European countries, especially in their tactics, in their building of left wing democratic alliances and in the issues raised in each case.

Two additional factors have contributed to the emergence of these new forms of struggle: the politicization and changing role of administration and the growing radicalization of petty-bourgeoisie and the peasantry. The changing role of administration refers primarily to the evolving challenge to the technocratic ideology of development through which the capitalist state ensures its welfare policies. This challenge comes from within and sometimes takes the form of a massive leftward politicization of broad sectors of the state personnel.[5] According to Buci-Glucksmann (1975), the administrative apparatus in southern Europe in the 1970s having become the target of multiform popular struggles, is less and was less able to resort to the ideological safety-screen of its role as 'neutral agent' above social classes. This was particularly evident in regional administrations controlled by the left, as in Emilia-Romagna and Toscana in Italy, where local development and management policies were often in conflict with those proposed by the central state (see, Zangheri, 1975).

There was also the growing radicalization of the petty-bourgeoisie and the peasantry. As in the case of administration, the recent experience of Fascism and military dictatorship on many occasions broke down the alliance between the bourgeoisie and petty-bourgeoisie and generated problems within the power bloc itself (Poulantzas, 1975). Because of their material conditions of existence, the petty-bourgeoisie and the peasantry actively participated in popular struggles, especially those relating to collective consumption and more general development issues. This new challenge to the traditional bourgeois/petty-bourgeois/peasant alliance found an important expression at the level of the state, around development policies concerning the production of space at the local and regional scales. The latter provided fresh possibilities for the formation of broader social alliances with the working class against monopoly capital and authoritarian statism.

These developments have contributed to the recent interest in social movements which coincides with efforts to develop a non-economistic view of capitalist societies able to articulate the logic of subsystems and mechanisms of stratification (Offe, 1976; Touraine, 1978; Cohen, 1982). The political mechanisms of stratification - applied mainly by the state and local authorities - involve a systematic logic that selects, creates and rewards certain kinds of interests and excludes others. At the same time, local authorities prevent universal participation in consensus formation by prestructuring the opportunities for specific strata to address local institutions or utilize their services. A three-tiered stratification results from this process:

a. Classes develop along the axis of market and production/reproduction relations. As we saw in previous chapters, classes are not derived simply from occupational structures, and there is no automatic link between economically-defined class position and the position taken in political struggle (Hirsch, 1981). Similarly, regional unevenness contributes to the formation of different kinds of classes in different regions and localities. As Cooke (1983) points out, the question is why some local classes develop solidarity and regional class consciousness, while others do not and whether class organization bears a clear relationship to particular regional labour markets.

b. Stratification also develops in the proliferation of certain interest groups across class lines generated by the political administrative (state and local) subsystem. Again, uneven regional development influences to a great extent the performance of different politico-administrative units.

c. Further stratification is associated with marginalized areas (backward regions, decayed urban centres, segregated neighbourhoods) marginalized groups (ethnic, cultural) and marginalized types of needs (child care, home health care, young non-qualified employment) constituted as such, i.e. as marginal, by the state, so that the possibility of being articulated _vis-à-vis_ the wider political and social system seems less and less possible.

These developments have contributed to a situation in which the state incurs a sizeable loss of legitimacy due to contradictions generated from its own interventionist policies, while at the same time the popular masses strengthen their position and alliances. This socio-spatial relation between the state and the regions has been described in Chapter 1 in terms of the relationship between 'regionalization' and 'regionalism'. Whenever exogenous intervention into regional/local life acquires a concrete spatial dimension, it may be called _regionalization_: a process designed 'from above' by the state, local authorities or capital, aimed at restructuring the spatial division of labour according to the changing needs of profitable accumulation. Whenever exogenous intervention penetrates the territorial domain of a region through a certain regionalization, a conflict may occur in the form of _regionalism_: a reaction of a specific local social group (or a certain social alliance) whose local interests are against such a regionalization. The conflict based on regionalism is likely to happen when the collective memory can provide evidence for proving _the social logic of the place_.

Historically, every social formation has been engaged in various forms of regionalism, as part of historical changes in the spatial division of labour. In the past, as the example of southern Europe clearly demonstrates (see Chapters 6 and 7), regionalist reactions were based on localism, on peasant struggles and/or ethnic cultural problems of peripheral minorities. The conjuncture today, however, has changed conditions that generate regionalist reactions by adding the critique against authoritarianism, clientelism and the technocratic ideology of economic progress. Thus contemporary regional consciousness in southern Europe combines cultural and economic elements with political practice, derived primarily from past political experience and recent social and spatial contradictions that uneven regional development creates.

My contention is that regional social movements in southern Europe are dependent on the interaction between regionalization and regionalism, in other words, the articulation between the state, capital (mainly multinational capital) and local community. This articulation and the conflict often associated with it must be viewed within the wider context of the structural crisis of the contemporary capitalist state and the rising popular consciousness around the political implications of regional problems. Thus, the problem is how a particular regionalization may generate a regionalist reaction and respectively how a regionalist strategy may be handled and integrated by a specific regionalization policy.

There are many cases, however, in which the articulation between regionalization and regionalism had not produced any significant regional mobilization. In these cases the proposed new spatial division of labour by the state or capital was either in agreement with local interest or it generated minor reactions which soon became integrated into the proposed restructuring policy. Moving now into the 'active mobilization side' and summarizing broadly current experience of regional mobilizations, three forms can be distinguished: historical socio-cultural differences with important current economic and political aspects; conflicts between two or more regional hegemonic blocs with minor popular participation; and conflicts between local dominated classes and a regional hegemonic bloc.

First, there still exist - expecially in Spain - old modes of articulation based on localism and strong ethnic and socio-cultural differences. Two such well-known cases here are the

Basque national-liberation movement and the autonomy movement in Catalonia. Less radical mobilizations exist also in Galicia, Andalusia and the Canary Islands. These regional social movements are standard references in any discussion on regionalism (together with Corsica, Wales, Scotland, Quebec and Walonia), and there is no need for an extended presentation here.[6] They have been active in Spain, since the beginning of this century, while their first institutional recognition for self-government was under the Republic in 1930. After Franco, the 'degree-law' of September 1977 and the new constitution of 1978 gave limited regional autonomy to the Basques and the Catalans, but ETA continues its armed struggle. As I mentioned in Chapter 7, various factions of ETA are demanding complete autonomy and separatism from both Spain and France. They have the vision of a free, socialist Basque country based on the alliance between petty-bourgeoisie, working class and the peasantry. In this respect, the military wing of ETA is in conflict with local élites (monopoly industrial and financial capital) which accept the proposal for limited autonomy from Madrid.[7]

Through these policies the Spanish state is trying to integrate the Basque movement and the same is true for the Catalans. Both movements, however, seem to be too dynamic and rich to disappear. Both are rooted in a living modern culture with a rich historical past and are enjoying an economic prosperity with the highest per capita income in the country. For these reasons, the Basques and the Catalans were the first to enjoy 'the state of Autonomies' introduced after referenda in October 1979. Then other less prosperous regions had been granted pre-autonomous status, including Galicia and Andalusia. This upsurge in regionalism, however, must not give the impression of the break up of Spain. In fact, such developments have been made possible without any danger to the general unity of the country due to the post-1950s deep transformation of both the economy and civil society (Giner and Sevilla, 1984).

A second form of regional mobilization is generated when a certain regionalization produces a <u>conflict between two regional hegemonic blocs</u> with minor popular participation. Regional hegemonic bloc is a notion derived from Gramsci and Poulantzas and consists of: (a) a specific system of economic exploitation and articulation among various forms of production; (b) a specific form of political and ideological cultural mediation towards dominated classes; and (c) a specific form of alliance between dominant classes under the hegemony of a regional bourgeoisie. In this respect the regional hegemonic bloc of the Mezzogiorno exercised its power towards dominated classes differently in comparison with northern Italy. The same is true in Spain for the case of industrialized north versus agricultural south. Thus, when the northern Italian bourgeoisie (using the Italian state) tried to change the system of exploitation in the south through industrialization, it faced strong reactions from local élites (Nocifora, 1978). Similar conflicts, but for different reasons, were generated between the Madrid and the Catalan regional hegemonic blocs.

Conflicts between regional hegemonic blocs were mediated through the state and in most cases took either a defensive form (i.e. against a particular regionalization which could threaten the local dominant position of the hegemonic bloc) or a demand form (i.e. demanding a certain intervention to increase or to change its competitiveness vis-à-vis other regional hegemonic blocs and/or the state). In these two cases the conflict and the breaking up of the rules occur within an organizational system, i.e. one which is characterized by roles and functions. The collective actors demand a different distribution of the resources within the organization, fight for a more efficient operation of the apparatus and finally clash with the power which imposes both the rules and the form of spatial division of labour. An example of such defensive mobilization was generated out of the antithesis between certain fractions of the Friuli-Venetian bourgeoisie and the Italian state.[8] After the alarm sounded by the 1966 flood, subsequent discussion led to broad-area planning proposals (see also, Arcangeli, 1982). The central and local government were immediately involved in subregional planning and this has threatened local bourgeois interest in polluting industries (in Mestre and Marghera) and in the tourist sector (historical city of Venice). These interests made a coalition to oppose planning proposals demanding a 'high authority', a special development agency as opposed to local council planning. They were also able to mobilize for a limited period of time the dormant Friuli-Venetian movement, a right wing middle class regional coalition. Behind this whole development, however, was an open conflict among different political forces including the centre-right coalition in government, the right wing industrial and tourist regional interests and the communist-led local councils.

Demand forms of mobilization include, among others, the Cretan case, in which the regional hegemonic bloc pushed the Greek state for additional economic subsidies to confront competition from EEC members, especially in agriculture and tourism. Among the major demands was the 'acceptance' by local police and state agencies of the use of low paid and non-declared seasonal foreign labour employed in agricultural fields and tourist resorts.

These conflicts among regional hegemonic blocs or between a regional hegemonic bloc and the state usually do not attract popular participation: they remain largely an internal affair of the power bloc. Furthermore, regional hegemonic blocs typically present their differences as differences of the popular masses. Problems such an unemployment, industrial investment, migration and means of collective consumption are examples used by capital and the state to divide the working class, the peasantry and segments of the petty-bourgeoisie of one region against those of another (see also, Garofoli, 1983).

An interesting example here was the case of Pylos, in Greece, where multinational capital found support from a regional 'movement' (Pylos-A), favouring its investment proposal. Pylos, a backward region in southern Peloponnesos, was a conflict area from 1974 to 1978 due to a proposal for the installation of a large port and industrial complex in the historical and environmentally important Navarino Bay. Our research showed that the multinational was supported mainly by petty-bourgeois landlords and right wing deputies, who were also able to mobilize local people under the prospects of a 'future industrial growth pole'.[9]

Under certain conditions, however, the interests of dominated classes can coincide temporarily with a fraction of the regional hegemonic bloc interests. This was the case, for example, in Sicily, where the popular masses made a temporary alliance with a fraction of the local bourgeoisie and the state to fight against the domination of the Mafia.[10] Another example was the Megara movement in Greece. In Megara, near Athens, workers, peasants and the petty-bourgeoisie made an alliance with local merchant élites against the installation of a foreign refinery and petrochemical complex in their area (Hadjimichalis, 1976).

Finally, the third form of mobilization takes place when regionalization generates a conflict between local dominated classes and a regional hegemonic bloc. In this case regionalist reaction is addressed either against local élites who have allied themselves with exogenous intervention forces, or directly against these exogenous forces. It is this form of mobilization, i.e. the rise of popular regional consciousness and the conflict between dominated and dominant classes around regional issues which I consider the prime example of a regional social movement. It is the case of a political movement, expressing a conflict through the rupture of the boundaries of the political system. It fights for the widening of political participation and struggles against the prevalence of the ruling interests within the representative system.

As in the case of regional hegemonic blocs, regional social movements can take both a defensive form (resisting, for example, a new spatial division of labour based on massive reallocation of jobs) or a demand form (e.g. asking for the improvement of means of collective consumption at the regional level). In both cases the movement is based on an anti-capital and anti-state social alliance among fractions of the petty-bourgeoisie, the working class and the peasantry.

Examples of defensive regional social movements include the Andalusian and the Messolongi (central Greece) mobilizations. In the first case, an initial petty-bouregeois/small peasant coalition made an alliance with the working class and formed a joint committee against the restructuring of agricultural production that would result in massive unemployment and greater concentration of land ownership (A. Titos Moreno and Rodriguez, 1979). In May and June 1979 'symbolic' land occupations took place by the radical Sindicato de Obreros de Cambo (SOC). Then, in August 1980 there was a ten day hunger strike involving hundreds of people to protect existing jobs and to obtain new ones. The SOC's main battle was against mechanization, while movement leaders were asking for improvements of the 'unequal trade relationships between Andalusia and the rest of Spain'.

In the second case, Mesolongi (a small historical town in the entrance of the Corinthian gulf), popular masses were successfully mobilized against the installation of a state petrochemical complex in their rich agricultural and fishing areas. Mobilization issues included pollution control, improvement of fishing and agricultural produce management and improvement of means of collective consumption at the regional level.[11]

In terms of the demand forms of regional social movements, good examples are the Emilia-Romagna Development Coalition, the Galician '<u>camaras agrarias</u>', the Pylos-B movement and the Grevena (northern Greece) mobilization. In Emilia-Romagna since 1962, in conjunction with the leftist regional administration, a development coalition was established based on four goals: democracy, decentralization, localized development and people's activation (Zangheri, 1975). The development coalition was founded on the broad anti-Fascist alliance of dominated classes (including the Christian-Democrats) established in the region immediately after World War II. The coalition was responsible for the region as a whole through a four year plan (Piano Programma) and new legislation (passed in 1974), while at the urban and neighbourhood levels more than 500 councils were established.[12]

In Galicia, since 1974 local working class people and radicalized peasants have been able to mobilize their neighbours against traditional power groups. Local political <u>caciques</u> (political bosses, mainly doctors) have always been prominent in Madrid, a tradition that was continued after Franco with the electoral success, nationally and within Galicia of Union Democratico de Centre (UDC). But since 1977 a new sense of regional consciousness has been forged by shared problems and cultural traditions, involving preservation of jobs, small-scale farming and fishing, the long history of out-migration, a common language and, above all, the realization of uneven regional development in Spain. Mass mobilization and political action, including a rally, led to the establishment of an independent slate for the '<u>camaras agrarias</u>' in the Spanish elections, based in part on the creation of agricultural cooperatives (Buechler and Buechler, 1978). As an opposition to Pylos-A 'movement', which supported the multinational investment, another social movement, Pylos-B, generated in the area during the end of the summer of 1976. This time the movement was against the project. Local people, with the help of outside radicals, organized a 'Committee for the Solution of Messinian Problems', demanding from the state an alternative development plan for the area. The committee came up with different development priorities, such as modern agro-industrial complexes under cooperative ownership, and tourist packages utilizing existing housing in depopulated villages.[13] Finally, the Grevena regional social movement in northern Greece has been a direct mobilization against regional backwardness.[14] Organized during the winter and spring of 1980, it started with a four day regional strike in which regional economic life was practically stopped. A social base composed of petty-bourgeois, workers and small peasants elected a coordinating committee, which presented a four-point declaration to the state; improvement of transportation, health and education facilities; improvement of the unequal trade relations of Grevena producers (especially peasants) with the rest of Greece and the EEC; organization of cooperative industrial units for the repatriation of 5-6,000 Grevenitians working in West Germany as _gastarbeiter_; and declaration of the Grevena prefecture as a first priority area for regional financial aid by the state and the EEC.

The typology of regional social movements is summarized in Figure 11.1, showing the distinction between reaction-non reaction to regionalization/ regionalism, dominant patterns of response to regionalization and social practices. The figure is only illustrative of the various categories of regional social movements discussed in the text and is not aimed to equalize such diverse cases as, for example, the long-established Catalonian movement with the short-lived reaction in Messolongi or Megara.

A primary conclusion from this brief and still incomplete discussion of empirical cases is that economic issues related to regional backwardness are not the singular determination or necessary sources for a regional mobilization. In many cases cultural and environmental issues as well as uneven distribution of means of collective consumption seem to be more important. This point has been addressed by other scholars (see also, Mormont, 1983; Cooke, 1983; Rokkan and Urwin, 1983; Slater, 1984) who also conclude that regional mobilizations are not confined to particular positions within the spatial division of labour. They can take place in core, semi-peripheral and peripheral regions alike, without necessarily an economic determination.

What these struggles have in common, however, is an ideological argument about territory. It is in the name of territory that some people set themselves up against others or against some social

Figure 11.1: Types of Regional Mobilizations

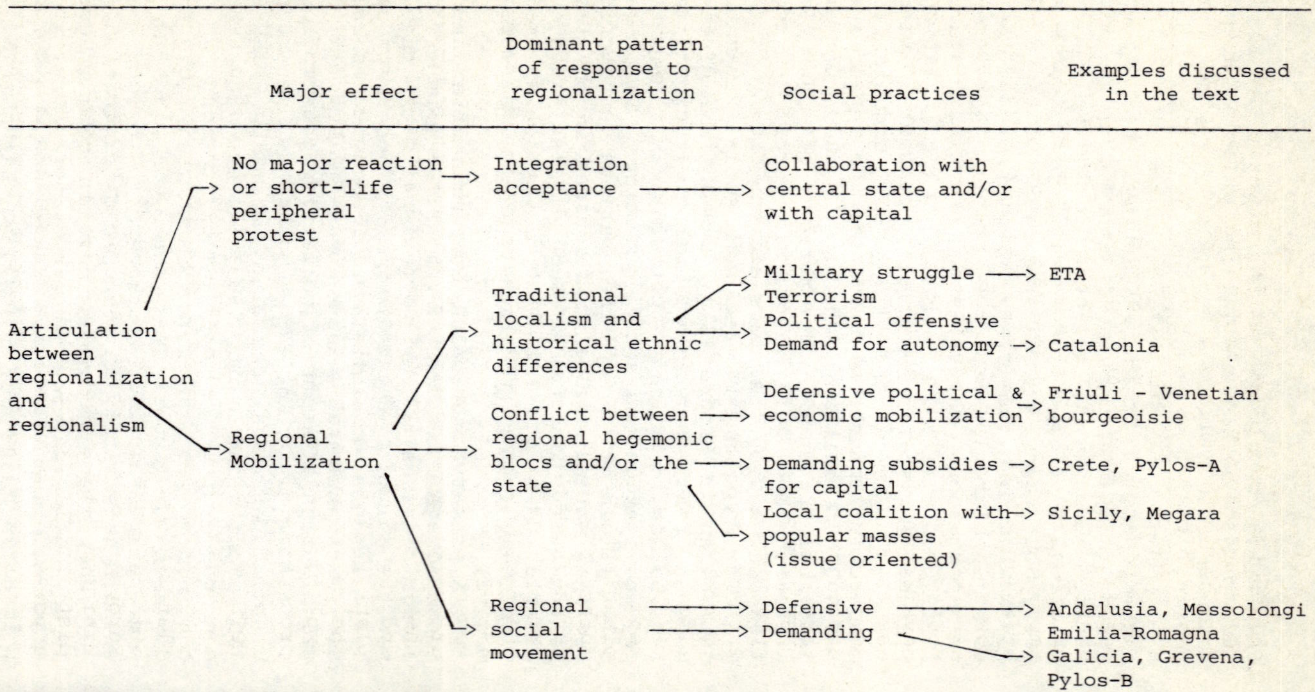

philosophy. Its effect is to bring about (or attempt to bring about) a wide mobilization based on the fact of belonging to the same membership space and geographical space.

Such a brief analysis of regional social movements in southern Europe cannot illustrate in detail either the planning inadequacies of capitalist state or the self-management potential on the part of regional popular masses. Further comparative empirical research is needed, until more elaborated statements deal with the issues. Present conditions, however, seem to point to the development of an alternative to institutional planning: social mobilization in relation to problems of uneven development at regional and local levels. Regional issues, alongside urban issues, that give rise to social movements (defensive or demanding) can be under the present conjuncture a primary tactical target for popular masses in southern Europe. Regional social movements can organize wide social strata around issues of regional backwardness and extend the initial mobilization for spatial issues to other sectors of social, political and cultural life.

It can always be argued that the ambiguity of regional social movements cannot secure their planning efficiency and their success *vis-à-vis* authoritarian statism and monopoly capital. Furthermore, as in any other social movement there exist possibilities for reactionary tendencies in the name of preservation of certain communal values, as for example defence *per se* of a traditional hegemonic bloc against a modernized one. But these are risks that the popular movement has to face when the objective is a democratic road to social change and finally to socialism. On this road, regional planning must formulate new alternatives and styles, the most important of which will be to adapt an explicit territorial focus and to open its procedures to popular participation.

NOTES

[1] See, Wallerstein (1983); Dulong (1978), and Slater (1984).

[2] A key source of power within southern European governments was always the post of the Minister of Public Works. The geographical distribution of public projects remains until today a powerful means for clientelism and patronage.

[3] Results from European Parliament elections

in 1984 and from national elections in Greece and Portugal show, however, a decline for the socialist parties which are in power. Contradictory policies promoted by these socialist governments were the main causes for such a decline.

[4] One should also note here the existence of important regional social movements in Portugal and France. See, Claude De Vos (1975) ; Dulong (1976), and Rokkan and Urwin (1983).

[5] During 1977 and 1978 a number of important strikes in state agencies illustrates this point: in Italy with post office personnel, the public health department and urban-based police departments; in Greece, with school teachers, bank personnel and the public electricity company. See, AVGI, 7 December 1978.

[6] See, among others, Mandel (1963); Lafont (1976); Cooke (1983), Rokkan and Urwin (1983).

[7] Since the referendum of March 1980 for autonomy - which ETA had unsuccessfully boycotted - there has been a growing radicalization of Basque people, including the Basque Nationalist Party. Contrary to expectations of the Madrid government separatist tendencies seem to be stronger than ever among all segments of the population. See, Le Monde, 10 August 1980.

[8] See, L'Unità, 10 April 1969.

[9] See, Hadjimichalis (1976); Hadjimichalis and Vaiou-Hadjimichalis (1979).

[10] This was particularly important during the 1950s when large landowners associated with the Mafia opposed land reform and managed, in fact, to do so rather effectively. A similar issue arose in the early 1970s, when local authorities, local business associations and trade unions mobilized to challenge the control by the Mafia of the tourist and building sectors.

[11] See newspaper, AVGI, 2 October 1979 (in Greek). The Messolongi mobilization became the theme of a documentary film (as in Megara) with a wide circulation.

[12] The PCI, however, did not avoid certain clientelistic practices at the local level in Bologna which damaged its wider political image. As, for example, during the events of March 1977 with massive street demonstrations, when certain CP members were helping the police to close down 'Radio Alice', an independent radical radio station; see, ANTI, no.110, p.36 (in Greek).

[13] This plan in fact was drafted by the Ministry of Planning and the Environment in 1980, but only the section of the physical planning for the two major towns of the area (Pylos and Methoni) proceeded for implementation.

[14] See newspaper, AVGI, 19 January 1980 and 21 January 1980, and newspaper, TO VIMA, 22 and 23 March 1980 (in Greek).

Chapter 12
STATE TERRITORY AND CLASS: CONCLUDING REMARKS AND SOME QUESTIONS AHEAD

In this work there are no conclusions as such, no easy answers to the questions raised. Nor do the results of my empirical investigation represent a new formal theory about uneven regional development. This was not the intention. My purpose has been to point to some of the sources of uneven regional development, taking specific historical examples from southern Europe. By considering regions and uneven regional development as the result of an endless struggle between regionalization and regionalism, over economic, political and cultural issues, I was able to propose a number of ideas about the relationship between social classes, social conflict, territory and the state.

Regions are historical products, not only in their physical materiality but in their socio-cultural meaning, in the role they play in the social organization of production and reproduction. In this respect they acquire a special position in the development of social formations around the Mediterranean, as does the process of uneven regional development. In the investigation of this process, I emphasized the contradiction of differentiation/equalization and the underlying tendency of differentially localized accumulation that results in the geographical transfer of value. This assumption may or may not be generally accepted. If it is accepted, however, and one is convinced that some capitalists in certain regions accumulate less surplus value than their workers produce, while others in other regions accumulate more, one must also accept that this can be effected mainly through a geographical transfer of value (inter or intra-sectoral) from the former to the latter.

I want to emphasize again the fact that the

production and reproduction of uneven regional development, and more precisely what today appears as the 'regional crisis' in southern Europe, exist over and above the economic transfer of surplus value or surplus labour from the direct producer in a region to the person who immediately receives it within the region or outside of it. The discussion and analysis of the agricultural sector in Spain, Italy and Greece aimed primarily at showing the strong interrelationship between economic, political and cultural/ideological factors in producing and reproducing uneven regional development.

Furthermore, by reconstructing the concrete geographical and social contexts in which uneven regional development takes place, I was able to demonstrate one of my basic hypotheses: that, although the GTV concerns tendencies which are possessed necessarily by capital by virtue of its structure, it operates through the mediation of, and closely depends upon, the actions of social agents, including the state. Thus, the bringing together of the elements of uneven regional development is mediated by various forms of struggle around issues such as the particular choice of technology and managerial practices, cultural inferiority and the forms of state intervention. No theory of regional development could be expected to know the nature and form of these contingencies in 'advance', purely on the basis of theoretical claims. In fact, of course, the real situation is always more complex, and in the brief discussion on agriculture, as well as in the regional mobilization studies presented in previous chapters, the political implications of uneven regional development became evident. Among the many political implications I distinguish here three key issues: first, the geographical transfer of value as a possible source of regional consciousness and potential focus for regional social movements; secondly, the role of political parties; and thirdly, the role of spatiality in contemporary capitalism.

Starting with the first - the political implications of the geographical transfer of value - and given the overall context of uneven regional development, the trade-off between one geographical area and another is far clearer. In cases of direct/indirect value and surplus labour transfers, or in cases of simple location shifts in employment, one region's gain is clearly another region's loss. This is manifest, not only in employment opportunities but additionally in wage and living standards differentials and in inequalities in the provision and quality of means of collective consumption. Capitalist competition and the state in southern Europe are taking advantage of historically established conditions of differentiated regional labour markets to reproduce uneven regional development and to discriminate against certain population groups in peripheral regions.

As a warning it should be added that this analysis in no way means that the only way for a group of workers or a wider population group to advance in a particular region is to decrease real wages and the general well-being of other groups in other regions. This in fact has often been promoted by capital in its divide and conquer strategy. As far as workers are concerned, however, it should go without saying that their task is to prevent discrimination and to reinforce their unity. This is of course today in southern Europe highly problematic, since the working class, lower middle strata and the peasantry are more than ever internally differentiated. The latter provided wide opportunities for capital to apply differentiated rates of exploitation from a social, spatial, racial and gender point of view. But that, in turn, merely reinforces the need for change in working class strategies. Working people in gaining and losing regions will both lose soon, due to continuous changes in the spatial division of labour (see Map 9.8 in Chapter 9). The understanding of this can open possibilities for certain class alliances while excluding others and build solidarity among people and movements at different scales, among different sectors and across regions.

It seems therefore that a regional consciousness which develops along certain political and cultural values and the geographical transfer of value can be of high value. From the regional social movements discussed in Chapter 11, only Andalusia and to some extent the Grevena movement seem to have developed such a conceptualization. The best historical example in southern Europe remains that of anarchist Catalonia during the civil war (see Chapter 6). Then, people became aware, among other things, of the operation of geographical transfer of value and attempted to control it through the introduction of common 'production brigades' in rural areas and in the cities. We cannot of course repeat today the Catalan experience. It seems, however, that a combined regional consciousness along the geographical transfer of value and along social, cultural and environmental issues can help to rethink the 'regional question' in a more concrete and politicized framework.

The second important issue is the role of political parties. Uneven regional development has been recognized by all political parties in southern Europe as a key problem in national development. Yet, their analyses and policies vary considerably, not only along a right-left dichotomy (which is not always clearly observable) but within the left itself. This became evident with the rise in power of southern socialist parties in the beginning of the 1980s. Despite their rhetoric, proposed 'socialist' regional policies did not differ much from an updated Keynsianism with neo-populist slogans (e.g. Greece) or even from neo-liberalism with a socialist coverage (e.g. France, Spain, Italy).

The crucial problem, however, with these socialist parties is their relation with mass movements and especially with regional social movements. As has already been noticed, all regional mobilizations are not progressive by definition, nor are they political movements. But without a reform of political economic steering mechanisms, the very social arena in which they operate and in which they seek to expand is endangered.

Yet one cannot simply call for the 'great alliance' between parties, unions and regional social movements. The history of social movements (regional and otherwise) has shown that, whenever a party, usually of the left, tries to 'guide' them - to bridge social demands with political targets - the social movement soon disintegrates and in many cases even dissolves.[1] The left has been unable to understand these new forms of social mobilization and was unprepared to accept their autonomy. With some exceptions the majority of leftist parties in southern Europe have not accepted their independent political status. The latter became a controversial issue under the political leadership of southern socialist parties. Their basic argument is that their strategy for 'change' incorporates automatically all social movements. This integrationist ideology builds upon a conceptualization of regional and local issues as being simply part of a national electoral strategy. The most illustrative example is PASOK in Greece, which calls itself a 'movement' and not a party. Under this coverage, certain types of social movements, such as the peace movement, the women's movement, certain regional mobilizations, the environmental movement and a few others '... can better operate within PASOK, rather than struggling independently out of it'.[2]

The situation in the mid 1980s is therefore different from the mass mobilizations of the 1960s and 1970s. On the one hand, development strategies of socialist parties now in power in southern Europe seem to focus on the needs of capital and to a 'strong nation-state' to compete internationally and to integrate any autonomous social movement in their strategy of 'change'. On the other, deepening economic crisis, high unemployment, regional capital restructuring and continuous cultural domination - to mention only a few hot issues - remain potential sources for mobilizations which attract, however, less and less the militant interest of regional populations. In this conjuncture, regional issues are again in the hands of central state bureaucracy (or the EEC) despite the fashionable decentralization rhetoric. With the absence of significant movement to the left of socialist (with the exception of PCI in Italy) the likelihood is that individualistic and corporastic demands and right wing opposition will be the main beneficiary from this situation.

Thus, while regional social movements in the 1960s and 1970s - as part of the wider radical mobilizations of the time - could help in developing planning alternatives against statist planning, today they have to struggle again for 'basics': for employment, for social services, for regional autonomy, or for the '*scala mobile*'. But, as history shows, social conflicts and struggles in Spain, Italy and Greece *simultaneously* generated new possibilities for the generation of new movements and *recreated obstacles* to any attempt by popular forces to appy their self-management policies. But the decisive task of any democratic leftist strategy is to grapple with this intensely contradictory situation, rather than to ignore the contradictions in the polemical defence of a specific choice.

Finally, the third relevant political issue is the growing importance of *spatiality* in contemporary southern Europe and the generation of conflicts around the social use and meaning of space. Indeed, I tried to present a statement about uneven regional development in Spain, Italy and Greece, placing the production of space and its role in the reproduction of social relations at the centre of the survival of modern capitalism. The process of social reproduction and spatiality are tightly intertwined in modern capitalism. Social space is not a simple 'container' of productive forces, but is actively linked to the reproduction of dominant social relations. Thus, the central importance of spatiality in modern capitalism is related to the changing conditions of accumulation and consumption, its new social meaning, the role of the state and rising popular spatial consciousness.

Given what has come before this final section, it should be clear that I do not aim at submerging class analysis, or at seeing space as an autonomous structure. My intention was to analyse spatial organization in direct conjunction with class analysis, to develop what Soja (1980) calls a 'socio-spatial dialectic'. In this respect an alternative regional analysis can help to raise social and territorial consciousness among dominated classes, formed with and not against class consciousness. This is an important point, distinguishing between the use of regional differentiation by the state and capital and the use of the same process by radical working class political organizations. The former use regional differences as a means to disorganize and to split the people into fragmented and antagonistic territorial units through a series of political interventions; the latter, acknowledging that exploitation is taking place <u>both</u> at the place of production and reproduction are able to see cultural, class, racial or gender distinctions dialectically related with their placement into particular 'places': in a constantly changing spatial division of labour, from global to local.

* * *

In the preceding section I have commented upon the political implications of uneven regional development and the changes in the spatial division of labour. Since the end of my research in the mid 1970s, these changes have become more evident. I summarized some of the new patterns of regional unevenness and the changing social stratification in Chapters 10 and 11. My basic hypothesis was that, during the period under study (1950-1975), agriculture was 'squeezed' to finance rapid industrial development and general economic growth, and this has led to severe social and spatial inequalities. Polarization patterns between industrial and rural regions were associated with extensive accumulation and 'economic miracles.' If, however, the sectoral focus on agriculture was a legitimate starting point to analyse at that time the process of uneven regional development, today it seems to be less so.

From the mid 1970s onwards, deep transformations in the southern European countryside associated with the wider economic crisis and capital restructuring changed the relationship between agriculture and the rest of the economy, civil society and the state. It is not any more the agricultural sector *per se* which calls for attention, but a complex set of productive activities dialectically related with *rural space*. Rural space was never homogeneous, and I tried to describe social and spatial differentiation patterns in Part II. But today new contradictions emerge, due to restructuring within agriculture, productive decentralization and flexible and irregular forms of employment in industry, tourism and the public sector. These new contradictions in rural space provided opportunities to quite a large number of small peasant producers, part-time workers and middle strata in the service sector (mainly spread over the hills, valleys and coastal plains) to avoid - at least up to the present day - the family disintegration and social marginalization and pauperization which have struck poor peasants in mountain regions and certain sections of the working class in big cities. This has been made possible thanks to the spreading of part-time work in several sectors and firms during the year and the combination of formal and informal activities to earn an income (Paci, 1979; Tsoukalas, 1984).

This new pattern of uneven development is not the direct product of capital's needs, nor does it depend for its reproduction on the realization of an average rate of profit. It is the outcome of changes in the spatial division of labour which gave to rural space a new content. Capitalist relations of production are mediated today in southern European countryside through new household forms of production which are extended far beyond subsistence and small peasant production. Furthermore, the greatest specificity of the southern European situation must be seen in the persistence (beyond all expectation by 'orthodox' Marxian interpretations) of a vast area inside the sphere of production dominated by simple commodity production and informal activities. Thus, class relations are based on social figures which - again following the orthodox dogma - should have disappeared altogether with the march of capitalist development.

A common interpretation of these characteristics is that of 'backwardness' and 'underdevelopment' (Mouzelis, 1978; Fuà, 1976). But this interpretation is unsatisfactory. It does not explain, in fact, why after a hundred years of industrial development (at least in Italy and Spain) southern European class structure includes such low numbers of wage labour and high numbers of self-employed, industrial petty bourgeoisie, peasant and artisan family production units, and marginal proletariat. It seems that the persistence of these social strata are so much a part of the specific development pattern of southern Europe, that it needs a different interpretation. A conceptualization therefore along a sectoral point of view (agriculture, industry or otherwise) and the use of official statistics only is not enough for an understanding of current changes in the southern European pattern of uneven regional development. Thus, the following five points can be seen as an opening of new questions to be asked which provoke simultaneously the re-evaluation of some basic concepts used in this work. After all, each ending should, in truth, be viewed as but a new beginning.

First, the new pattern of uneven regional development seems to combine widening and deepening of regional inequalities, making traditional distinctions such as urban-rural, north-south or core-periphery highly problematic. As I mentioned in Chapter 7, inequalities are developing now among and within regions depending on a system of localized social relations and increasingly fragmented regional labour markets. Within the context of a prolonged economic crisis, certain intermediate regions (defined in terms of rate of growth of GRP, population density, activity mixture and degree of accessibility) started to grow rapidly. Their sectoral and geographical composition of their productive system is more flexible and diversified compared with old industrialized regions or marginal mountainous areas. In these regions the relative development of infrastructure and communications, certain regional incentives to capital, the existence of good irrigated land, beaches and monument sites have permitted - due to the lack of strict land use regulations - the coexistence of medium and small industrial firms, along with tourist facilities and intensive agricultural production. Through efficient use of local resources, certain local entrepreneurs have been able to go through the recession and the present economic crisis which hit particularly hard old industrialized regions.

Secondly, the present position of southern European economics in the international and particularly EEC division of labour caused a progressive specialization of their productive structure towards certain industrial products (or industrial activities) for which small firms are especially functional; their agriculture production combines traditional Mediterranean crops with new short-cycle, soft products for which small peasant producers can compete with large capitalist farms; and their monuments, history and beaches attract mass tourism which moves now away from large complexes to small-scale tourist resorts or to preserved vernacular settlements. This particular specialization, however, is strongly linked to two factors: the uncertainty of demand, always dependent on the rapid changes in fashion, taste and consumption patterns; and the maturity of the existent technologies which are moving towards important labour-saving innovations and more diffused patterns of production. Changes in these two factors will cause considerable unevenness among and within southern European regions.

Thirdly, a specific characteristic which calls for attention in the present conjuncture is the importance of the informal sector. Informal activities were historically always present in southern European economies, but today must be analysed in the light of the new mode of articulation between capitalism (defined principally by wage-labour) and petty commodity production (defined principally by non-wage relations and self-employment). This new articulation is neither restricted to agriculture alone, nor can be associated with conditions of backwardness. In fact, as many studies have shown, the development of small and very small industrial and tourist firms in rural regions is the flexible response to new demands and to direct needs for restructuring and surviving through a prolonged economic crisis. Thus, new forms of informal activities are spreading, such as subcontracting, work by piece at home, room letting, operating bars and restaurants parallel to traditional agricultural work. These activities are not considered as marginal or outside capitalist relations of production, but on the contrary are integral components of the new pattern of uneven development, which provides fresh room for accumulation. In fact, it became clear that the surplus generated within the informal sector (in value terms or as surplus labour) is impossible to be retained by its 'members'. At a less abstract level, this implied that the existing relations between small firms and larger capitalist enterprises were themselves the major obstacles to any sustained economic growth. These small firms are contributing more through various forms of transfers to the process of capital accumulation *outside* its own limited sphere of production than to local growth as was expected.

Fourthly, the organization of the regional labour market in those intermediate regions is an important component of the new development pattern (see also, Chapters 7 and 8). The labour force is more diversified, combining effectively formal and informal employment opportunities, less unionized and in certain cases lacking a syndicalist tradition. The cost of the labour hour *per se* is not cheaper compared with old industrial centres, but part-time jobs in different sectors and firms are more widespread and acceptable to local workers (Ginatempo, 1985). Thus, small firms are able to escape both full year payrolls and social security payments. The picture is completed by seasonal work (in agriculture and tourism), work by-piece at home, room-letting and a variety of irregular jobs in the industrial and tourist sectors. In many cases the industrial experience of return migrants is playing a key role as they act as mediators between large foreign firms and small local subcontracting firms or with individual women working at home (Hadjimichalis and Vaiou-Hadjimichalis, 1985).

This range of activities is not a one-worker's operation. It is accommodated within the *household*, whose members (taken as a group) have formal, full-time employment, engaged in informal, home-based activities, in seasonal work and even black work (Vinay, 1985). Its place within the extended family usually provides the household with land property necessary for most activities mentioned above (for production as well as for access to loans). In this respect, the specific demographic and cultural characteristics of rural southern Europe provide the preconditions for the spread of this development pattern. One can exclude, however, the spread of this type of development in regions which are already very degraded or marginal (e.g. mountain regions), or towards certain large urban centres, where there exists neither economic nor social, demographic and cultural possibilities of combining this complex range of activities (Hadjimichalis, 1984).

Moreover, with rising inflation and reduced formal employment, the household secures the availability of goods and services that would otherwise have to be purchased in the market (Mingione, 1985). These services go far beyond housekeeping proper, to include food production, house-building and repairs, education and socialization of family members and assistance to the ill and old. The household then acquires again special importance as a unit of production, consumption and reproduction. This new role of the household is one of the consequences of the current employment crisis in southern Europe in the context of which unemployment and non-employment increase without a parallel decrease in the cost of reproduction of labour force. Thus, working commitments within the family unit not only tend to increase, taking up nearly all available leisure time, but also are very discriminatory between sexes and the various age groups. In this respect, old patriarchical and authoritarian relations are reproduced within the family, with women being always at the bottom of the hierarchy (Vaiou, 1985).

To sum up, the specificity of southern European class structure and its relation to territory (at least for the three countries studied) must be linked to this new development pattern. The peculiarity of southern Europe is that this new pattern provides an 'assistance' to a growing number of the population as well as an outlet for capital accumulation in a period of severe economic crisis. This is realized through the maintenance of a 'worker-peasant-entrepreneur' context tied to informal activities, small size production units and household forms of production and reproduction. An explanation, therefore, of regional development in southern Europe cannot be accurate unless it includes a detailed analysis of the articulation of these production and reproduction processes as they are manifested today in the regions around the Mediterranean.

In this conjuncture the left is presented with acute questions for which it is ill-equipped to produce meaningful answers, both at the theoretical and practical political levels. At a time of crisis, when the principal contradictions are neither clearly nor universally defined, the left has to challenge its own hierarchies of human activity and incorporate forms of exploitation and sources of collective consciousness and struggle that extend far beyond the point of production. Such an approach, however, cannot rest on individual efforts. It needs to be related to a mass movement committed to social and spatial transformation. For the left, southern Europe today seems to present opportunities, as well as dangerous complications for such mass movement efforts. An analysis of these processes may help to clarify the components and the ways in which alternatives can be formulated. But it cannot resolve the problem. Its solution is political and lies in concrete struggles and social movements which acknowledge explicitly the spatiality of modern capitalism as a tied relationship between class, territory and the state. Not as a substitute for the general class struggle but as a specified orientation within it.

NOTES

[1] Perhaps an extreme but illustrative case is the peace movement in Greece, where three organizations exist, each controlled by a different party.

[2] See the interview by A. Papandreou in TA NEA, 15 May 1984 (in Greek).

REFERENCES

Aglietta, M. (1976) Régulation et Crises du Capitalisme, Calman-Levy, Paris

Aiello, C. (1972) 'L'Aménagement de l'Exploitation Agricole dans les Zones d'Exode Rural: Italie', in L. van Nieuwenhuije (ed.), Emigration et Agriculture dans le Bassin Méditerranéen, Mouton, Paris

Alaiz, F. (1937) 'Por una Economia Solidaria entre el Campo y la Cuidad', CNT-FAI, Barcelona

Alonso, W. (1967) 'Urban and Regional Imbalances in Economic Development', Economic Development and Cultural Change, no.17

Alonso, W. and Friedmann, J. (ed.) (1975) Regional Policy: Readings in Theory and Applications, The MIT Press (2nd edition), Cambridge

Alvater, E. (1973) 'Notes on Some Problems of State Interventionalism', in Kapitalistate, Issue 1/2, pp.18-31

Amin, S. (1973) Le Développement Inégal, Editions Minuit, Paris

Amin, S. (1974) Accumulation on a World Scale (2 vols.), Monthly Review Press, New York

Amin, S. (1977) Imperialism and Unequal Development, Monthly Review Press, New York

Amin, S. and Vergopoulos, C. (1974) La Question Paysanne et le Capitalisme, Anthropos, Paris

Amsden, J. (1979) 'Industrial Collectivization under Workers' Control: Catalonia 1963-69', Antipode, Vol. 10-11, no.3, pp.99-112

Arcangeli, F., Bozzaga, C., and Groglio, S. (1980) 'Patterns of Peripheral Development in Italian Regions 1964-77', paper presented to the 19th European Congress of the RSA, London (mimeo)

Arcangeli, F. (1982) 'Regional and Subregional Planning in Italy: An Evaluation of Current Practice and Some Proposals for the Future', in

Hudson, R. and Lewis, J. (eds.), Regional Planning in Europe, Pion, London

Bagnasco, A. (1977) Tre Italie, Il Mulino, Bologna

Bagnasco, A. (1980) 'Development based on Small Industry in Italy: An Intermediate Economic and Social Model'

Banco de Bilbao (1973-1978) La Renta Nacíonal de España, Bilbao

Baran, P. (1957) The Political Economy of Growth, Monthly Review Press, New York

Baran, P. and Sweezy, P. (1970) Monopoly Capital, Penguin Books

Barbagallo, C. (1948) La Questione Meridionale, Milan

Barkan, O.L. (1954) 'La Méditerranée de Fernand Braudel, Vue d'Istamboul', Annales ESC, IX, 2

Barquero, A.V. (1983) 'Endogenous Industrialization in Rural Areas of Spain', paper presented at the International Conference, Valencia (December)

Bartra, R. (1975) 'Peasants and Political Power in Mexico: A Theoretical Model', Latin American Perspectives, Vol.II, no.2

Barzanti, S. (1965) The Underdeveloped Areas within the Common Market, Princeton University Press

Bauer, O. (1924) Die Nationalität Fragen und die Sozialdemokratie, Vienna

Belli, A. and Flora, A. (1978) Le Forbice Bisogni Sociali/Ricorse Finanziarie in una Grande Città dell'Italia Meridionale, ATTI, 1. Seminario Internazionale del Workshop Area Mediterranea, Milano

Benedito, Y. and Fuentes, E. (1978) La Rénovation Pédagogique à Barcelone, Xme Colloque de l'Association Internationale des Sociologues de Langue Francaise, Toulouse

Bettelheim, C. (1966) Problèmes Théoriques et Pratiques de la Planification, Maspéro, Paris

Blalock, H.N. (1972) Social Statistics, McGraw-Hill, New York

Bleitrach, D. and Chenu, A. (1979) L'Usine et la Vie: Luttes Régionales, Marseille et Fos, Maspéro, Paris

Boccara, P. (1971) Traité d'Economie Marxiste: Le Capitalisme Monopoliste de l'Etat, Editions Sociales, Paris

Boggs, C. (1976) Gramsci's Marxism, Pluto Press, London

Bolloten, B. (1968) The Grand Camouflage: The Spanish Civil War and Revolution, Praeger, New York

Borja, J. (1976) 'Les Mouvements Sociaux Urbains en Espagne', Espaces et Sociétés, no.19

Borts, G.H. and Stein, H.L. (1962) 'Regional Growth and Maturity in the USA: A Study of Regional Structural Change', in Needleman, L. (ed.) Regional Analysis, Penguin Books

Boudeville, J.R. (1964) L'Espace et les Pôles de Croissance, PUF, Paris

Boudeville, J.R. (1966) Problems of Regional Planning, Edinburgh University Press

Bowler, I. (1976) 'Recent Developments in the Agricultural Policy of the EEC', Geography, 61, pp.28-30

Braudel, F. (1972) The Mediterranean and the Mediterranean World in the Age of Philip II (2 Vols.) Harper, New York

Brehau, G. (1943) The Spanish Labyrinth, Cambridge University Press, London

Breitbart, M.M. (1979) 'Anarchist Decentralization in Rural Spain 1936-39: The Integration of Community and Environment', Antipode, Vol.10, no.3/Vol.11, no.1

Brenner, R. (1977) 'The Origins of Capitalist Development: A Critique of Neo-Smithian Marxism', New Left Review, no.104, pp.25-92

Bretonnière, B. (1976) 'A Propos des Régions en Italie: Histoire, Portée et Limites d'une Réforme', Espace et Sociétés, no.19

Browett, J. (1984) 'On the Necessity and Inevitability of Uneven Spatial Development under Capitalism', International Journal of Urban and Regional Research, Vol.8, no.2, pp.155-76

Brown, M.B. (1974) The Economics of Imperialism, Penguin Books

Buci-Glucksmann, C. (1975) Gramsci et l'Etat, Fayard, Paris

Buechler, H. and Buechler, J.M. (1978) 'Social Class, Conflict and Unequal Development in Spanish Galicia', The Review of Radical Political Economics, Vol.19, no.3

Bukharin, N. (1973) Imperialism and World Economy, Monthly Review Press, New York

Caballero, A. (1979) 'La Teoria de la Renta Absoluta, Grenta de Transformacíon o de Monopolio?' Agricultura y Sociedad, no.12

Camilleri, A. Nadal, E., Posada, J.L., Sumpsi, J.M., and Tío, C. (1977) La Exploitacíon Agraria Familiar, Serie Estudios, Ministerio de Agricultura, Madrid

Campbell, J.K. (1974) Honour, Family and Patronage in Rural Greece, Oxford University Press

Carney, J., Hudson, K. and Lewis, J. (eds.) (1980) Regions in Crisis, Croom Helm, Kent

Carr, R. and Fusi, J.P. (1981) Spain: Dictatorship to Democracy, G. Allen and Unwin, London

Carassi, E. (1975) 'Nouveaux politiques aux Régions Italiennes' Espace et Société, no. 17, pp.160-81

Castells, M. (1972) La Question Urbaine, Maspero, Paris

Castells, M. (1983) The City and the Grassroots, Edward Arnold, London

Castells, M. and Godard, F. (1974) Monopolville: L'Entreprise, l'Etat et l'Urbain, Mouton, Paris

Caviria, M. (1976) Ecologismo y Ordenacíon del Territorio en España, Edicusa, Madrid

Ceccareli, P. et al. (1978) Processus Socio-Economiques, Institutions, Systèmes Politiques et l'Organization du Territoire dans le Bassin Méditerranéen, 1. Seminario Internazionale del Workshop Area Mediterranea, Milano

Centro de Estudios Territoriales y Ambientales (CETA) (1975) Planificazione e Conflitto nelle Grandi Aree Metropolitane: Il Caso della Grecia, Italia, Spagna, Portogallo, Seminario Internazionale, Madrid

Chayanov, A.V. (1966) The Theory of Peasant Economy, Aldine, Chicago

Chenery, H.B. (1968) 'Development Policies for Southern Italy' Quarterly Journal of Economics, Vol.76

Cherpakov, V. (1969) State Monopoly Capitalism, Progress Publishers, Moscow

Cinanni, P. (1972) Emigrazione e Imperialismo, Roma

Clark, G. and Dear, M. (1981) 'The State in Capitalism and the Capitalist State', in M. Dear and A.J. Scott (eds.), Urbanization and Urban Planning in Capitalist Societies, Methuen

Clark, G. and Dear, M. (1984) State Apparatus, G. Allen and Unwin, Boston

Cockburn, C. (1977) The Local State, Pluto Press, London

Cohen, J. (1982) 'Between Crisis Management and Social Movements: The Place of Institutional Reforms', TELOS, no.52, pp.21-40

Cooke, P. (1982) 'Class Interests, Regional Restructuring and State Formation in Wales', International Journal of Urban and Regional Research, Vol. 6, pp.187-203

Cooke, R. (1983) Theories of Planning and Spatial

Development, Hutchinson, London
Coraggio, J.L. (1975) 'Polarization, Development and Integration', in Kuklinski, A. (ed.), Regional Development and Planning: International Perspectives, Leyden
Coriat, B. (1978) Un site en Méditerranée: Le Complexe Industrial-portuaire de Fos-sur-Mer, 10 Seminario Internazionale del Workshop Area Mediterranea, Milano
Cottone, E. (1972) Riorganizzazione Capitalista e Lotta di Classe nelle Campagne, Roma
Crisafulli, F. (1975) 'Poli Industriali, Territorio e Classe Operaia nel Sud Meridione', Città e Campagna, no.4-6

Damette, F. (1980) 'The Regional Framework of Monopoly Exploitation: New Trends', in Carney, J., Hudson, R. and Lewis J. (eds.), Regions in Crisis, Croom Helm, Kent
Darwent, D. (1969) Growth Poles and Growth Centers in Regional Planning - A Review', Environment and Planning, Vol. 1
Dedousopoulos, L. (1985) 'Simple Commodity Production, Capitalism and Commercial Capital', POLITIS, pp.3-11 (in Greek)
De Janvry, A. (1981) The Agrarian Question and Reformism in Latin America, The Johns Hopkins University Press
Della Seta, P. (1978) 'Notes on Urban Struggles in Italy', International Journal of Urban and Regional Research, Vol.2, no.2
Dematteis, G. (1983) 'Deconcentrazione Metropolitana, Crescita Periferica e Ripopolamento di Aree Marginali: Il Caso dell'Italia', in Cencini, C., Dematteis, G. and Menegatti, B. (eds.)
De Vos, C. (1975) Etat et Mouvements Regionaux, IXme Colloque de l'Association Internationale des Sociologues de Langue Francaise, Menton
De Vroey, M. (1981) 'Value, Production and Exchange', in The Value Controversy, Verso, London
Donolo, C. (1972) 'Sviluppo Inequale e Disegregazione Sociale: Note per l'Analisi delle Classe nel Meridione', Quaderni Piacentini, no.47
Downie, J. (1958) The Competitive Process, Duckworth, London
Dulong, R. (1976) 'La Crise du Rapport Etat/Société Locale vue au Travers de la Politique Régionale', in Poulantzas, N. (ed.) La Crise de

l'Etat, PUF, Paris

Dulong, R. (1978) Les Régions, l'Etat et la Société Locale, PUF, Paris

Duncan, S. and Goodwin, M. (1982) 'The Local State and Restructuring Social Relations: Theory and Practice', International Journal of Urban and Regional Research, Vol. 6, no. 2. pp.157-86

Eliou, M. (1978) 'The Geographical Distribution of Educational Opportunities', The Greek Review of Social Research, no. 28

Elson, D. (ed.) (1979) Value: The Representation of Labour in Capitalism, CSE Books, London

Emmanuel, A. (1969) L'Echange Inégal, Maspéro, Paris

Emmanuel, A. (1974) 'Myths of Development versus Myths of Underdevelopment', New Left Review, no.85

Emmanuel, A. (1977) 'State and Transition to Socialism', New Left Review, no.113-14

Emmerij, L. (1982) 'New Conservatism, New Realism and All That', in EADI Bulletin, 1/82, pp.3-15

Eudes, D. (1970) Les Kapétanios, Arthène-Fayard, Paris

Eurostat (1979) Regional Statistics 1970-1977: Main Regional Indicators, Brussels

Filias, V. (1972) 'Restructuring of Agricultural Enterprises Affected by Emigration: Greece', in van Nieuwenhuije, C. (ed.), Emigration et Agriculture dans le Bassin Méditerranéen, Mouton, Paris

Folin, M. (1981) 'The Production of the General Conditions of Production and the Role of the State', in Harloe, M., and Lebas, E. (eds.), City, Class and Capital, E. Arnold, London

Frank, A.G. (1967) Capitalism and Underdevelopment in Latin America, Monthly Review Press, New York

Frank, A.G. (1979) Dependent Accumulation and Underdevelopment, Monthly Review Press, New York

Friedman, A. (1977) Industry and Labour, Macmillan, London

Friedmann, J. (1966) Regional Development Policy: A Case Study of Venezuela, The MIT Press

Friedmann, J. (1973) Urbanization, Planning and National Development, Sage Publications, Beverly Hills

Friedmann, J. and Douglas, M. (1975) 'Agropolitan Development: Towards a New Strategy for Regional Development in Asia', United Nations

Centre for Regional Development, 1975, pp.333-87

Friedmann, J. and Weaver, C. (1979) Territory and Function: The Evolution of Regional Planning, University of California, Berkeley

Fuà, G. (1976) Occupazione e Capacità Prodottiva: La Realità Italiana, Bologna

Garcia-Ramon, D. (1978) 'The Shaping of a Rural Anarchist Landscape: Contributions from Spanish Anarchist Theory', Antipode, Vol.10, no.3/Vol.11, no.1

Garcia-Ramon, D. (1984) 'Limiting Factors in the Transformation of Peasant Agriculture in the Mediterranean: A Case Study from the Catalan Coast', Department of Geography, Universitat Autonoma de Barcelona (mimeo)

Garcia-Ramon, D. (1982) 'Agricultural Change in an Industrializing Area: The Case of the Tarragona Area', paper presented at the Conference on National and Regional Development in the Mediterranean Basin, University of Durham, (April)

Garofoli, G. (1974) Regional Inequality and the Labour Market, paper presented at the 'Réunion des Chercheurs Européens sur la Question Régionale', Paris

Garofoli, G. (1976) Une Analyse Critique de la Politique de Réequilibre Régional en Italie: Le Cas du Mezzogiorno, paper presented in 'Le Rencontre Européen sur les Problèmes Régionaux', Barcelone

Garofoli, G. (1983) Industrializzazione Diffusa in Lombardia, F. Angeli, Milano

Germidis, D. and Delivanis, M. (1975) Industrialization, Employment and Income Distribution in Greece, OECD, Paris

Giddens, A. (1981) A Contemporary Critique of Historical Materialism, Macmillan, London

Ginatempo, N. (1985) 'Social Reproduction and the Structure of Marginal Areas in Southern Italy: Some Remarks on the Role of the Family in the Present Crisis', International Journal of Urban and Regional Research, Vol.9, no.1, pp.99-110

Giner, S. (1982) 'Political Economy, Legitimation and the State in Southern Europe', paper presented at the Conference National and Regional Development in the Mediterranean Basin, University of Durham (April)

Giner, S. and Sevilla, E. (1984) 'Spain: From Corporatism to Corporatism', in Williams, A.

(ed.) Southern Europe Transformed, Harper and Row, London
Goldsmith, W.W. (1978) 'Marxism and Regional Policy: An Introduction', The Review of Radical Political Economics, Vol.10, no.3, pp.13-17
Goodman, D. and Redclift, M. (1981) From Peasant to Proletarian, Basil Blackwell, Oxford
Gore, C. (1984) Regions in Question: Space Development Theory and Regional Policy, Methuen, London
Gramsci, A. (1975) Selections from Political Writings 1910-20, International Publishers, New York
Granados, V. (1984) 'Small Firms and Rural Industrialization in Spain: Some Results from an OECD Project', in Hudson, R. (ed.), Small Firms and Regional Development, Institute for Transport, Tourism and Regional Development, Copenhagen (Publication No.39)
Graziani, A. (1975) 'Aspetti Strutturali dell'Economia Italiana nell'Ultimo Decennio', in Graziani, A. (ed.), Crisi et Ristrutturazione nell'Economia Italiana, Einandi, Torino
Graziano, L. (1978) 'Center-periphery Relations and the Italian Crisis: The Problem of Clientelism' in Tarrow, S. et al. (eds.), Territorial Politics in Industrial Nations
Greek Ministry of Agriculture (1974) Agricultural Statistics, Athens
Griffin, K. (1979) The Political Economy of Agrarian Change, Macmillan, London
Gruppi, L. (1977) Il Compromeso Storico, Riuniti, Roma
Guillot, M.F. and Roux, Y. (1977) Les Mouvements Nationaux Régionaux, Documment de Travail, AISLF, Paris

Habermas, J. (1973) Theory and Practice, Beacon Press, Boston
Habermas, J. (1975) Legitimation Crisis, Beacon Press, Boston
Hadjimichalis, C. (1980) The Geographical Transfer of Value: A Comparative Analysis of Regional Development in Southern Europe, PhD, Graduate School of Architecture and Urban Planning, UCLA
Hadjimichalis, C. (1983) 'Regional Crisis, the State and Regional Social Movements in Southern Europe' in Seers, D. and Öström, K. (eds.), The Crises of the European Regions, Macmillan, London

Hadjimichalis, C. (1984) 'The Geographical Transfer of Value: Notes on the Spatiality of Capitalism', *Society and Space*, Vol.2, pp.329-45

Hadjimichalis, C. and Vaiou-Hadjimichalis, D. (1979) *Regional Development and Industrialization*, Exandas Publishers, Athens (in Greek)

Hadjimichalis, V. and Vaiou-Hadjimichalis, D. (1980) 'Penetration of Multinational Capital into Backward Regions: A Policy Analysis in Greece', *Mediterranean Peoples*, no.10

Hadjimichalis, C. and Vaiou, D. (1985) 'Changing Regional Labour Markets and Forms of Social Reproduction in Greece', paper presented at the Conference on *Spatial Structures and Social Processes*, in Lesvos, Greece, (August/September)

Hansen, E.C. (1977) *Rural Catalonia under the Franco Regime*, Cambridge University Press

Hansen, M.N. (ed.) (1972) *Growth Centers in Regional Economic Development*, The Free Press, New York

Harloe, M. (ed.) (1978) *Captive Cities*, John Wiley and Sons, New York

Harris, R.L. (1978) 'Marxism and the Agrarian Question in Latin America', *Latin American Perspectives*, Vol. V, no.4., pp.2-26

Harvey, D. (1973) *Social Justice and the City*, The Johns Hopkins University Press

Harvey, D. (1974) 'Population, Resources and the Ideology of Science', *Economic Geography*, Vol.50, no.3

Harvey, D. (1975) 'The Geography of Capitalist Accumulation: A Reconstruction of the Marxist Theory', *Antipode*, Vol.4, no.2

Harvey, D. (1982) *The Limits to Capital*, Basil Blackwell, Oxford

Hastaoglou, V. (1982) 'Le Rapport Ville-Campagne chez Gramsci', *City and Region*, no.3, pp.7-33 (in Greek)

Hechter, M. (1975) *Internal Colonialism*, University of California Press, Berkeley

Hill, B. (1984) *An Agricultural Policy for the EEC*, Methuen, London

Hirsch, J. (1981) 'The Apparatus of the State, the Reproduction of Capital and Urban Conflicts', in Dear, M. and Scott, A.J. (eds.) *Urbanization and Urban Planning in Capitalist Societies*, Methuen, London

Hirschmann, A. (1958) *The Strategy of Economic Development*, Yale

Hoare, Q. and Navell-Smith, G. (eds.) (1971)

Selections from the Prison Notebooks of Antonio Gramsci, Lawrence, London
Hodgson, G. (1981) 'Critique of Wright' in Steeman, I. et al. (eds.), The Value Controversy, Verso, London
Holland, S. (1971) 'Regional Under-development in a Developed Economy: The Italian Case', Regional Studies, Vol.5
Holland, S. (1975) The Socialist Challenge, Quartet Books, London
Holland, S. (1976) Capital versus the Regions, Macmillan, London
Holloway, J. and Picciotto, S. (eds.) (1978) State and Capital, E. Arnold, London
Hoover, E.M. (1967) The Location of Economic Activity, McGraw-Hill, New York
Hoselitz, B. (1960) Sociological Factors in Economic Development, The Free Press, Glencoe
Hudson, R. (1983) 'Capital Accumulation and Chemicals Production in Western Europe in the Postwar Period', Environment and Planning, Vol.15, pp.105-22
Hudson, R. and Lewis, J.R. (1984) 'Capital Accumulation: The Industrialization of Southern Europe?' in Williams, A. (ed.), Southern Europe Transformed, Harper and Row, Cambridge
Hymer, S. (1972) 'The Multinational Corporation and the Law of Uneven Development', in Bhagwati, J. (ed.), Economics and the World Order from the 1970s to the 1990s, Macmillan, London

Illich, I. (1980) 'The New Frontier for Arrogance: Colonization of the Informal Sector', Development, Vol.XXII, no.2-3, pp.96-101
Indovina, F. (1977) 'Les Forces Sociales et l'Usage de la Région', Cahiers de l'ACPOL, no.2
Ingrao, P. (1976) 'Des Régions pour Unir', Critica Marxista, no.65
Instituto Nationale de Estadistica Agraria (INEA) (1978) Annuario dell'Agricultura Italiana, Vol.XXXI, Mulino, Bologna
Isard, W. (1960) Methods of Regional Analysis: An Introduction to Regional Science, MIT Press
Iturra, R. (1980) 'Strategies in the Domestic Organization of Production in Rural Galicia', Department of Anthropology, University of Cambridge (mimeo)

Jaggi, M, Müller, and Schmidt, S. (1977) Red Bologna, Writers and Readers, London
Jesus, C.M. (1970) El Espacio Regional en el

Dessarrollo Economico de España: Reflexiones Criticas, Ediciones Zero, Madrid
Johnson, B.F. (ed.) (1967) Agricultural Development and Economic Growth, Cornell University Press, Ithace
Jones, A.R. (1984) 'Agriculture: Organization, Reform and the EEC', in A. Williams (ed.), Southern Europe Transformed, Harper and Row, London
Judez, L., Garcia-Velazquez, A. et al. (1979) Analyse Comparative des Structures Agricoles au Niveau Régional de l'Espagne, la France, la Grèce, l'Italie et le Portugal devant les Perspectives d'Elargissement de la CEE, Institut Agronomique Méditerranéen de Montpellier, Montpellier

Kafkalas, G. (1981) 'Regions in the Greek Economy' City and Region, no.2, pp.7-38 (in Greek)
Kaldor, N. (1975) 'Why are Regional Policies Necessary?', in Regional Politik und Agrarpolitik in Europa, Heft 22, Berlin
Kalogirou, N. (1980) 'Organization and Production of Space in Illegal Settlements', Technika Chronika no.2/1980 pp.51-8 (in Greek)
Kaplan, T. (1976) Anarchists of Andalusia, Princeton University Press
Karagiorgas, D. (1980) The Economic Functions of the State, Papazisis, Athens (in Greek)
Karapostolis, V. (1984) Consumer Behaviour in Greek Society 1960-75 National Centre for Social Research, Athens (in Greek)
Kavouriaris, M. (1974) 'Some Thoughts on Migration' in Nicolinakos, M. (ed.), Economic Development and Migration in Greece, Kalvos, Athens (in Greek)
Kay, G. (1975) Development and Underdevelopment: A Marxist Analysis Macmillan, London
Kayser, B. (1970) Manpower Movements and Labour Markets (Italy, Spain, Turkey, Greece, Portugal), OECD, Paris
Kayser, B., Péchoux, C. and Sivignon, J. (1971) Exode Rural et Attraction Urbaine en Grèce, National Centre for Social Research, Athens
Kidron, M. (1974) Capitalism and Theory, Pluto Press, London
King, R. (1973) Land Reform: The Italian Experience, Butterworth, London
Kiray, M.B. (1973) Social Stratification and Development in the Mediterranean Basin, Mouton, Paris
Koshimura, S. (1975) Theory of Capitalist

Reproduction and Accumulation, DPG Publishing Company, Ontario
Kuklinski, A. (ed.) (1979) Growth Poles and Growth Centers in Regional Planning, Mouton, Paris
Kuznets, S. (1955) 'Economic Growth and Income Inequality', American Economic Review, no.49
Kuznets, S. (1956) Modern Economic Growth, Yale University Press, New Haven

Laclau, E. (1977) Politics and Ideology in Marxist Theory, Verso, London
Laclau, E. and Mouffe, S. (1981) 'Socialist Strategy: Where Next?' Marxism Today, no.25, pp.25-32
Lacoste, Y. (1972) 'La Géographie' La Philosophie des Sciences Sociales, Hachette, Paris
Läpple, D. and van Hoogstraten, P. (1980) 'Remarks on the Spatial Structure of Capitalist Development, the Case of the Netherlands', in Carney, J., Hudson, R. and Lewis, J. (eds.), Regions in Crisis, Croom Helm, Kent
Lasuen, J.R. (1969) 'On Growth Poles', Urban Studies, Vol.6
Lebas, E. (1978) 'Regional Policy Research: Some Theoretical and Methodological Problems', in Harloe, M. (ed.), Captive Cities, John Wiley, Chichester
Ledrut, R. (1975) Sur le Mode d'Existence Historique des Nations, IXme Colloque de l'Association Internationale des Sociologues de Langue Francaise, Toulouse
Lee, R. (1977) 'Regional Relations and Economic Structure in the EEC', in D. Massey and Batey, P.M. (eds.) Alternative Frameworks for Analysis, London
Lee, R. and Ogden, P.E. (eds.) (1976) Economy and Society in the EEC: Spatial Perspectives, Saxon, London
Lefeber, L.C. (1966) Location and Regional Planning, Centre of Planning and Economic Research, Athens
Lefebvre, H. (1968) Le Droit à la Ville/Espace et Politique, Anthropos, Paris
Lefebvre, H. (1971) La Pensée Marxiste et la Ville, Gasterman, Paris
Lefebvre, H. (1974) La Production de l'Espace, Anthropos, Paris
Lenin, V.I. (ed.) (1956) The Development of Capitalism in Russia, Foreign Language Publishers, Moscow
Lenin, V.I. (1973) Imperialism: The Highest Stage of

Capitalism, Progress Publishers, Moscow
Lewis, R. (1973) 'Agricultural Taxation and Intersectoral Resource Transfers', 'Food Research Institute Studies in Agricultural Economics, Trade and Development, Vol.12, no.2, Stanford University
Libertini, L. (1973) 'Si puo parlare di Proletarizzazione?', Rinascità, no.13
Liossatos, P. (1979) Unequal Exchange and Regional Disparities, paper presented at the 26th North American Meeting of the Regional Science Association, Los Angeles
Lipietz, A. (1975) 'Structuration de l'Espace, Problème Foncier et Aménagement du Territoire', Environment and Planning, Vol.7
Lipietz, A. (1977) Le Capital et son Espace, Maspéro, Paris
Lipietz, A. (1981) 'The So-called Transformation Problem Revised', Journal of Economic Theory, Vol.26, pp.59-88
Lipietz, A. (1983a) 'Le Débat sur la Valeur: Bilan Partiel et Perspectives Partiales', research paper 8326, CEPREMAP, Paris
Lipietz, A. (1983b) 'Sur les Fordismes Péripheriques de l'Europe du Sud', City and Region, no.7, pp.137-58
Little, M.D., Scitovsky, T. and Scott, M. (1970) Industry and Trade in Some Developing Countries, London
Lojkine, J. (1977) Le Marxisme, l'Etat et la Question Urbaine, PUF, Paris
Lopez, R.S. (1973) 'The Trade of Medieval Europe: The South', in Rostan and Rich (eds.), Cambridge Economic History of Europe, Cambridge University Press, London
Lorenzo, C.M. (1969) Les Anarchistes Espagnols et le Pouvoir (1868-1969), Editions de Seuil, Paris
Loucopoulos, Ch. (1979) 'The Problem of Immigrants' Integration' in Nicolinakos, M. (ed.) Economic Development and Migration in Greece, Kalvos, Athens (in Greek)
Louloudis, L. (1984) 'Agricultural Modernization and Restructuring Policy', International Conference on Agriculture and Political Economy, Athens (mimeo)
Lutz, V. (1962) Italy: A Study in Economic Development, London

Magnaghi, A. and Perelli, B. (1978)'Ristrutturazione e Diffusione Territoriale del Ciclo Produttivo: Formazione della Fabbrica Diffusa" in Italia',

ATTI - 1. Seminario Internazionale del Workshop Area Mediterranea, Milano

Malefakis, E.E. (1970) Agrarian Reform and Peasant Revolution in Spain, Yale University Press, New Haven

Mandel, E. (1963) 'The Dialectic of Class and Region in Belgium', New Left Review, no.20

Mandel, E. (1975) Late Capitalism, New Left Books, London

Mandel, E. (1976) 'Capitalism and Regional Disparities', Southeast Economy and Society, Vol.1, no.1

Mandi, P. (1983) 'Factors Distorting Realistic Evaluation in Development Studies' EADI Bulletin, no.1, pp.23-8

Marelli, E. (1983) 'Empirical Estimation of Intersectoral and Interregional Transfers of Surplus Value: The Case of Italy', Journal of Regional Science, Vol.23, no.1, pp.49-70

Markusen, A. (1979a) 'Regionalism and the Capitalist State: The Case of the United States', Kapitalistate, Vol.7, Winter

Markusen, A. (1979b) 'Capital Accumulation, Capital Mobility and Regional Planning', paper presented at the Annual Meeting of the American Planning Association, Baltimore (mimeo)

Markusen, A. (1983) 'Profit Cycles, Oligopoly and Regional Transformation', working paper 397, Institute of Urban and Regional Development, Berkeley

Marx, K. (1867/1975) Kapital (3 Volumes), Modern Publishers, New York

Massey, D.B. (1977) 'Towards a Critique of Industrial Location Theory', in Peet, R. (ed.), Radical Geography, Methuen, London

Massey, D.B. (1978a) 'Regionalism: Some Current Issues', Capital and Class, no.6

Massey, D.B. (1978b) 'In What Sense a Regional Problem?', paper presented to the Regional Studies Association Conference, Glasgow

Massey, D.B. and Batey, P.W. (eds.) (1977) Alternative Frameworks for Analysis, London Papers in Regional Science, Pion, London

McNeill, W.H. (1978) The Metamorphosis of Greece since World War I, Chicago

Meek, R.L. (1956) Studies in the Labor Theory of Value, Monthly Review Press, New York

Melucci, A. (1981) 'Ten Hypotheses for the Analysis of New Movements', in Pinto, D. (ed.) Contemporary Italian Sociology, Cambridge University Press, London

Miliband, R. (1974) The State in Capitalist Society, Quarter Books, London

Mingione, E. (1976) Divisione Territoriale del Lavoro e Sviluppo Capitalistico, Convegno Internazionale su i Problemi Sociali dello Sviluppo, Roma

Mingione, E. (1977) 'Theoretical Elements for a Marxist Analysis of Urban Development', in Harloe, M. (ed.), Captive Cities, John Wiley, London

Mingione, E. (1978) 'Capitalist Crisis, Neo-Dualism and Marginalization', International Journal of Urban and Regional Research, Vol.2, no.2

Mingione, E. (1981) Social Conflict and the City, Basil Blackwell, Oxford

Mingione, E. (1985) 'Social Reproduction of the Surplus Labour Force: The Case of Southern Italy' in Redclift, N. and Mingione E. (eds.), Beyond Employment, Basil Blackwell, Oxford

Ministerio de Agricultura (1977) Annuario de Estadistica Agraria, Secretaria General Technica, Madrid

Mintz, P. (1970) L'Autogestion dans l'Espagne Révolutionnaire, Belitaste, Paris

Moreno, A.T. and Rodriguez, J.J. (1979) Crisis Economica y Empleo en Andalucia, Ministerio de Agricultura, Madrid

Morishima, M. (1979) Marx's Economics: A Dual Theory of Value and Growth, Cambridge University Press, Cambridge

Mormont, M. (1983) 'The Emergence of Rural Struggles and their Ideological Effects', International Journal of Urban and Regional Research, Vol.7, no.4, pp.559-78

Moseley, M.J. (1974) Growth Centres in Spatial Planning, Pergamon, London

Moskof, K. (1974) The National and Social Conscience in Greece: 1830-1909 The Merchant Ideology, Athens (in Greek)

Mottura, G. and Pugliese, E. (1975) Agricoltura, Mezzogiorno e Mercato de Lavoro, Mulino, Bologna

Mottura, G. (1980) 'Appunti sulla Questione Agraria in Italia tra gli Anni '30 a gli anni '50', Agricoltura e Societa, II

Mouffe, C. (ed.) (1979) Gramsci and Marxist Theory, Routledge and Kegan Paul, London

Mouzelis, N. (1978) Modern Greece: Facets of Underdevelopment, Macmillan, London

Murray, R. (1971) 'The Internationalization of Capital and the Nation State', New Left Review,

no. 67
Myrdal, G. (1957) Economic Theory and Underdeveloped Regions, London

Nicolinakos, M. (1973) Politische Ekonomie der Gastarbeiterfrage: Migration und Kapitalismus, Rowohlt Tachenbuch Verlag, Reinbek
Nocifora, E. (1978) 'The Southern Question', International Journal of Urban and Regional Research, Vol.2, no.2
Novacco, D. (1976) La Questione Meridionale, Torino

O'Connor, J. (1973) The Fiscal Crisis of the State, St. Martin Press, New York
OECD (1971) Agricultural Policies in Spain, Paris
OECD (1972) Agricultural Policies in Italy, Paris
OECD (1973) Salient Features of Regional Development in Spain, Paris
OECD (1974) Agricultural Policies in Greece, Paris
OECD (1979) Re-Appraisal of Regional Policies in OECD Countries, Paris
Offe, C. (1974) 'Structural Problems of the Capitalist State', in German Political Studies, Vol.1, Ed. K. von Beume, Sage, London
Offe, C. (1976) 'An Analysis of Late Capitalist Societies', International Journal of Politics, Fall, pp.67-85
Ohlin, B. (1933) Interregional and International Trade, Harvard

Paci, M. (1978) 'Il Mercato del Lavoro dall'Unità d'Italia da Oggi', in Tranfaglia N. (ed.) Il Mondo Contemporaneo, Firenze
Paci, M. (1979) 'Strutture Sociale en Società Italiana' Archives Européenes de Sociologie, XX. pp.49-55
Paelinck, J. (1965) 'La Théorie du Développement Régional Polarisé', Cahiers de l'ISEA, Vol.15
Paggi, L. (1979) 'Gramsci's General Theory of Marxism', in Mouffe, C. (ed.), Gramsci and Marxist Theory, Routledge and Kegan Paul, London
Palloix, C. (1972) 'The Question of Unequal Exchange: A Critique of Political Economy' Bulletin of the Conference of Socialist Economists, Vol.2
Palloix, C. (1975) L'Internationalisation du Capital: Eléments Critiques, Maspéro, Paris
Palombara, La., J. (1966) Italy: The Politics of Planning, Syracuse University Press
Panitsidis, G. (1983) Marxism and the Agrarian

Question in Greece, Modern Times, Athens (in Greek)
Papayannakis, M. (1976) 'Crise et Construction Hégémonique d'un Nouvel Ordre Economique International en Méditerranée', The Greek Review of Social Research, no.26-7
Papayannakis, M. (1978) L'Agriculture et le Développement des Etats Méditerranéens, Université Libre de Bruxelles, Brussels
Pastore, G. (1966) Linee di una Nuova Politica Industriale per il Mezzogiorno, Palermo
Perez-Agote, F. et al. (1978) Le Mouvement Basque, IXme Colloque de l'Association Internationale des Sociologues de Langue Francaise, Toulouse
Perloff, H.S. and Dunn, E.S. et al. (1960) Regions, Resources and Economic Growth, Johns Hopkins University Press
Perloff, H.S. and Wingo, L. (1961) 'Natural Resource Endowment and Regional Economic Growth', in Alonso, W. and Friedmann, J. (eds.), Regional Policy Readings in Theory and Application, The MIT Press, Cambridge
Perroux, F. (1952) 'Les Pôles de Développement', Economie Appliquée, Vol.1, no.4
Perroux, F. (1955) 'Note sur la Notion des Pôles de Croissance', Economie Appliquée, Vol.3, no.1-2
Petras, J. (1984) 'The Rise and Decline of the Southern European Socialist Parties', New Left Review, no.146, pp.37-52
Petrella, R. (ed.) (1972) Le Développement Régional en Europe, Mouton, Paris
Pickvance, L.G. (1972) Urban Sociology: Critical Essays, Tavistock, London
Pinnaró, G. and Pugliese E. (1985) 'Informatization and Social Resistance: The Case of Naples', in Redclift, N. and Mingione, E. (eds.) Beyond Employment, Blackwell, London
Poulantzas, N. (1968) Classes Sociales et Pouvoir Politique de l'Etat Capitaliste, Maspéro, Paris
Poulantzas, N. (1974a) Les Classes Sociales dans le Capitalisme Aujourd'hui, Editions de Seuil, Paris
Poulantzas, N. (1974b) 'On Social Classes', New Left Review, no.92
Poulantzas, N. (1975) La Crise des Dictatures: Portugal, Grèce, Espagne, Maspéro, Paris
Poulantzas, N. (ed.) (1976) La Crise de l'Etat, PUF, Paris
Poulantzas, N. (1978) State, Power, Socialism, Verso, London
Preteceille, E. (1981) 'Collective Consumption, the

State and the Crisis of Capitalist Society', in Harloe, M. and Lebas, E. (eds.), City, Class and Capital, E. Arnold, London

Psichogios, D. (1984) 'The Surplus Extraction from Agriculture', Economy and the Agricultural Sector, proceedings of International Conference, Athens

Psiroukis, N. (1976) History of Modern Greece, Athens (in Greek)

Pugliese, E. (1979) 'Evoluzione della Struttura di Classe nel Mezzogiorno', in Graziani, L. and Pugliese, E. (eds.), Investimenti e Disoccupazione nel Mezzogiorno, Il Mulino, Bologna

Pugliese, E. (1982) 'Farm Workers in Italy: Agricultural Working Class, Landless Peasants, or Clients of the Welfare State?', paper presented at the Conference on National and Regional Development in the Mediterranean Basin, University of Durham (April)

Rey, P.P. (1973) Les Alliances de Classes, Maspéro, Paris

Richardson, H.W. (1969) Regional Economics, Praeger, New York

Richardson, H.W. (1971) 'Regional Development Policy in Spain', Urban Studies, no.8

Richardson, H.W. (1975) Regional Development Policy and Planning in Spain, Saxon, London

Rodriguez, F. (1972) 'La Politica Regional de los Planes Espanoles de Desarrollo', Boletin de Estudios Economicos, no.27

Rokkan, S. and Urwin, D.W. (1983) Economy Territory, Identity, Sage, London

Rostow, W. (1962) The Stages of Economic Growth - A Non-Communist Manifesto, Cambridge University Press

Sack, R.D. (1974) 'The Spatial Separatist Theme in Geography', Economic Geography, no.59, pp.1-19

Sakelis, M. (1985) Savings Behaviour of Greek Peasants, Agricultural Bank of Greece, Athens (in Greek)

Salvatori, M. (1979) 'Gramsci and the PCI: Two Conceptions of Hegemony', in Mouffe, C. (ed.) Gramsci and Marxist Theory, Routledge and Kegan Paul, London

Saraceno, P. (1959) Iniziativa Privata e Azione Publica nei Piani di Sviluppo Economico, Milano

Saraceno, P. (1973) 'I Progetti Speciali', Nord e

Sud, no.218
Sasson, J. (1977) L'Industrializzazione del Mediterraneo: Movimenti di Mano d'Opera e Capitali, Instituto Affari Internazioniali, Milano
Sassoon, A.S. (1980) Gramsci's Politics, Croom Helm, Kent
Saunders, P. (1981) Social Theory and the Urban Question, Hutchinson, London
Sauvy, A. (1970) Théorie Générale de la Population, PUF, Paris
Sayer, A. (1977) 'The Law of Value and Uneven Development: Some Problems and Possibilities for Analysis', paper presented to the Lanehead Workshop, Regional Theory Group (mimeo)
Schickele, R. (1968) Agriculture and Economic Development, Praeger, London
Schultz, T.W. (1965 Economic Crisis in World Agriculture, University of Michigan Press, Ann Arbor
Scitovsky, T. (1954) 'Two Concepts of External Economics', Journal of Political Economy, Vol.6
Scott, A.J. (1980) The Urban Land Nexus and the State, Pion, London
Secchi, B. (1977) 'Central and Peripheral Regions in a Process of Economic Development: The Italian Case', in Massey, D. and Batey, P.W. (eds.), Alternative Frameworks of Analysis, Pion, London
Secchi, B. et al. (1978) La Nouvelle Qualité des Relations de Dépendence/Intégration dans le Bassin Méditerranéen, 10 Seminario Internazionale del Workshop Area Mediterranea, Milano
Seers, D. (1969) 'The Meaning of Development', International Development Review, December, no.2-3
Sereni, E. (1956) Vecchio e Nuove nelle Campagne Italiane, Roma
Sereni, E. (1962) La Politica Agraria del Regime Fascista 1918-1948, Lezioni e Testimonianze, Milano
Servolin, C. (1972) 'L'Absorption de l'Agriculture dans le Mode de Production Capitaliste', in Colin A. (ed.), L'Univers Politique de Paysans, Paris
Shaikh, A. (1980) 'Foreign Trade and the Law of Value', Science and Society, Vol.12, pp.281-302
Shaikh, A. (1981) 'The Poverty of Algebra', in The Value Controversy, Verso, London
Slater, D. (1975) 'Underdevelopment and Spatial

Inequality', Progress in Planning, Vol.4, Part 2

Slater, D. (1984) 'The State and Issues of Regional Analysis in Latin America', in Walton, J. (ed.), Capital and Labour in an Urbanized World, Sage, London

Skouteri-Didaskalou, N. (1984) The Woman Question: An Anthropological Perspective, Politis, Athens (in Greek)

Soja, E. (1980) 'The Socio-Spatial Dialectic', Annals of the Association of American Geographers, Vol.10, no.2

Soja, E. (1981) 'A Materialist Interpretation of Spatiality', paper presented at the Workshop on the Geographical Transfer of Value, Canberra

Soja, E. (1983) 'Territorial Idealism and the Political Economy of Regional Development', City and Region, no.6, pp.55-74

Soja, E. and Hadjimichalis, C. (1979) 'Between Geographical Materialism and Spatial Fetishism: Some Observations on the Development of Marxist Spatial Analysis', Antipode, Vol.11, no.3

Sraffa, P. (1960) Production of Commodities by Means of Commodities, Cambridge University Press, MA

Steedman, I. (1977) Marx after Sraffa, New Left Books, London

Stöhr, W. and Taylor, D. (1981) Development from Above or Below?, John Wiley, Chichester

Stöhr, W. and Todtling, F. (1977) 'Spatial Equity - Some Antitheses to Current Regional Development Doctrine', papers and proceedings of the Regional Science Association, no.38, pp.33-53

Stoianovich, I. (1971) 'Material Foundations of Preindustrial Civilization in the Balkans', Journal of Social History, Vol.IV, no.3

Sunkel, O. (1976) 'Transnational Capitalism and National Disintegration in Latin America', in Meier, G.M. (ed.), Leading Issues in Economic Development, Oxford University Press, New York

Svoronos, N. (1972) Histoire de la Grèce Moderne, PUF, Paris

Sweezy, P. (1956) The Theory of Capitalist Development, Monthly Review Press, New York

Sylos Labini, P. (1965) 'La Politica Economica del Fascismo e la Crisi del 1929', Nord e Sud, no.70

Tamames, R. (1978) Estructura Economica de Espana, Alianza Universidad Textos, Madrid

Tarrago, M. (1976) 'Modèle de Développement et Politique Urbaine en Espagne', Espaces et

Sociétés, no.19

Teixidor, L.F.A. and Hebbert, M. (19 2) 'Regional Planning in Spain and the Transition to Democracy', in Hudson, R. and Lewis, J. (eds.), *Regional Planning in Europe*, Pion, London

Thompson, P. (1983) *The Nature of Work*, Macmillan, London

Tiebout, C. (1956) 'Exports and Regional Economic Growth', *Journal of Political Economy*, Vol.64

Tinbergen, J. (1964) 'Un Modèle de la Dispersion Géographique de l'Activité Economique', *Revue d'Economie Politique*, 1.74

Toft Jensen, H., (1982) 'The Role of the State in Regional Development, Planning and Implementation: The Case of Denmark', in Hudson, R. and Lewis, J. (eds.), *Regional Planning in Europe*, Pion, London

Touraine, A. (1978) *La Voix et le Regard*, Seuil, Paris

Treves, R. (1965) *Le Classi Sociali in Italia*, Torino

Tsoukalas, C. (1975) *Dépendence et Reproduction: Le Role Social des Appareils Scolaires en Grèce*, Paris

Tsoukalas, C. (1981) *Social Development and the State: The Organization of Public Space in Greece*, Themelio, Athens (in Greek)

Tsoukalis, C. (1984) 'The Miracle of Small and Medium Size Firms in Greece', *ANTI*, no.213, p.36 (in Greek)

Tsoukalis, L. (1981) *The European Community and its Mediterranean Enlargement*, G. Allen and Unwin, London

Urry, J. (1981a) *The Anatomy of Capitalist Societies*, Macmillan, London

Urry, J. (1981b) 'Localities, Regions and Social Class', *International Journal of Urban and Regional Research*, Vol.5, pp.455-73

Vaiou, D. (1985) 'Space-Gender Relations in the Production of the Built Environment', *National Technical University*, May, Athens (mimeo)

Vaitzos, C. (1974) 'Employment Effects of Foreign Direct Investments', in Edwards, E. (ed.), *Employment in Developing Nations*, New York

Van Nieuwenhuije, C.A. (ed.) (1972) *Emigration et Agriculture dans le Bassin Méditerranéen*, Mouton, Paris

Vergopoulos, C. (1975) *The Peasant Question in Greece: The Problem of Social Integration of*

Agriculture, Exandas, Athens (in Greek)
Vicens, V.J. (1969) An Economic History of Spain, Princeton University Press
Vilar, P. (1962) La Catalogne dans l'Espagne Moderne, Seuil, Paris
Vilar, P. (1985) 'Marxist History, a History in the Making: Towards a Dialogue with Althusser', New Left Review, no.76
Vinay, P. (1985) 'Family Life Cycle and the Informal Economy in Central Italy', International Journal of Urban and Regional Research, Vol.9, no.1, pp.82-97

Walker, R. (1978) 'Two Sources of Uneven Development under Advanced Capitalism: Spatial Differentiation and Capital Mobility', The Review of Radical Political Economics, Vol.10, pp.28-37
Wallerstein, I. (1974) The Modern World System: Capitalist Agriculture and the Origins of the European World Economy in the Sixteenth Century, Academic Press, New York
Wallerstein, I. (1976) 'A World-System Perspective on the Social Sciences', British Journal of Sociology, Vol.XXVII
Wallerstein, I. (1983) Historical Capitalism, Verso, London
Ward, B. (1964) Problems of Greek Regional Development, Centre of Economic Research and Planning, Athens
Warren, B. (1973) 'Imperialism and Capitalist Industrialization' New Left Review, no.81
Weaver, C. (1982) 'The Limits of Economism: Towards a Political Approach to Regional Development and Planning', in Hudson, R. and Lewis, J.R. (eds.), Regional Planning in Europe, Pion, London, pp.184-202
Webber, M.J. (1982) 'Agglomeration and the Regional Question', Antipode, no.14, pp.1-11
Webber, M.J. and Foot, S.P.H. (1984) 'The Measurement of Unequal Exchange', Environment and Planning A, Vol.16, pp.927-47
Whitman, M. (1967) 'International and Interregional Payments Adjustments: A Synthetic View', Princeton Studies in International Finance, no.19
Williams, A. (1984) (ed.) Southern Europe Transformed, Harper and Row, London
Williamson, J.G. (1965) 'Regional Inequality and the Process of National Development', Economic Development and Cultural Change, Vol.11, no.13

World Bank (1966) The Economic Development of Spain, The Johns Hopkins University Press

Wright, E.O. (1979) Class, Crisis and the State, Verso, London

Zalider, K. (1972) 'Capitalisme, Pollution et Mouvements Sociaux: Le Cas de Bilbao', Espaces et Sociétés, no.9

Zangheri, B. (1975) 'Red Bologna: Example or Exception?', Interview in Red Bologna, Jaggi, M. et al., London

NAME INDEX

PLACE INDEX

SUBJECT INDEX